Women Becoming Mathematicians

Women Becoming Mathematicians
Creating a Professional Identity in
Post–World War II America

Margaret A.M. Murray

The MIT Press
Cambridge, Massachusetts
London, England

This book was set in Sabon by Achorn Graphic Services, Inc., on the Miles System, and was printed and bound in the United States of America.

Library of Congress Cataloging-in-Publication Data

Murray, Margaret Anne Marie.
  Women becoming mathematicians: creating a professional identity in post–World War II America / Margaret A.M. Murray.
    p.   cm.
  Includes bibliographical references and index.
  ISBN 0-262-13369-5 (HC: alk. paper)
  1. Women mathematicians—United States. I. Title.

QA27.5.M88 2000
510′.82′0973—dc21

                                            99-08716

To the memory of
Laura Mayer (Ph.D., Yale, 1985; 1957–1997)
and
Nadine Kowalsky (Ph.D., Chicago, 1994; 1966–1996)

Contents

Preface

In recent years considerable attention has been devoted to the problem of underrepresentation of women in the fields of science and engineering. These fields have tended to attract and retain women in inverse proportion to the degree to which their subject matter is perceived to be difficult, abstract, and inaccessible. Typically, physics and many of the engineering disciplines attract and retain the fewest women, both numerically and proportionally, while the biological and behavioral sciences seem to attract the most. Mathematics has tended to fall somewhere in between, perhaps toward the lower end of the spectrum.[1]

Mathematics occupies a unique position among scientific and technical fields. It provides a logical and quantitative framework for the empirical sciences and at the same time has its own independent subject matter. Because the objects of mathematical study are concepts that have been abstracted from common notions of counting, measurement, geometry, and relationship, mathematics is not an empirical science in any ordinary sense of the word. So it is perhaps not surprising that even educated lay persons regard mathematics as difficult, abstract, and inaccessible. But one of the advantages of mathematics as a subject of study for women is precisely this: mathematical objects are portable and can be carried about in the mind, without need for any special equipment. In principle, mathematical inquiry can be carried out anywhere at all.

But meaningful mathematical inquiry does not occur in isolation; it relies on communication between mathematicians and on a community consensus regarding the accuracy, validity, and significance of new discoveries in the field. Similarly, the dissemination of mathematical knowledge is a social and cultural phenomenon, involving teachers, students, textbooks, and other means of instruction. The historical development of mathematics has been fueled by the resources, support, and encouragement of the larger society: business leaders, scientists, educators, politicians, policymakers. In short, mathematics depends on social relationships—both within its own community and in the larger communities of which it is a part. For women to make meaningful contributions to mathematics, they must establish productive relationships within these communities.

How, then, do women become mathematicians? How do they find satisfying work and earn respect and remuneration in a field that is largely defined and dominated by men? Clearly, the answers to these questions depend on myriad intellectual, social, political, cultural, and personal factors. The history of women in mathematics, as in other male-dominated professions, has followed a cyclic pattern of inclusion and exclusion (Graham 1978). In the United States, women mathematicians have always been in the minority, but they have during two distinct eras gained a substantial foothold in the profession. During the years 1910 to 1939, women earned roughly 14 percent of the mathematics Ph.D.'s awarded in this country, and most of those women went on to substantial careers in research, teaching, or administration. Since 1980, women have earned over 17 percent of the mathematics doctorates and have been more visible and active participants in mathematical life than have any previous generation of women mathematicians in history.

What of the intervening years, then? World War II was the signal event that changed the complexion of women's involvement in the American mathematical community. Through the participation of unprecedented numbers of mathematicians in war-related research, mathematics came to be seen as indispensable to national security. During the postwar years, higher education in general, and mathematics in particular, underwent unprecedented expansion. As mathematics grew in power and prestige, it became particularly attractive to men—men who might otherwise have pursued careers in law, medicine, or another of the sciences. At the same time, the prevailing cultural message for women was that victory in war meant a return to domesticity and femininity: a woman's place was in the home, rearing children and providing a comfortable haven for her husband.

While the numbers of women earning doctorates in mathematics remained essentially flat in the 1940s and 1950s, the numbers of men earning these degrees increased more than threefold. The net result was that women were gradually squeezed out of their place in American mathematics. So devastating was the reversal of women's earlier gains in the mathematical community that it took another two decades to recover the ground that had been lost.

During the postwar years, as men joined the mathematical profession in record numbers, a picture emerged of the life course of the ideal mathematician that was well suited to a *man's* life circumstances and particularly to those of a married man whose wife did no work outside the home. I refer to this as *the myth of the mathematical life course,* or simply, *the myth,* and it became the dominant model for the unfolding of mathematical careers in the decades following World War II.

While the economic and political conditions of the forties and fifties made it possible for increasing numbers of men to lead careers in conformity to the myth, its intensification in the postwar years cannot be explained solely in these terms. In her classic study *Men and Women of the Corporation,* Rosabeth Moss Kanter (1977) argues that when women's preponderance in a social group is reduced below a certain critical mass, they assume a token status. She asserts that "tokens are, ironically, both highly visible as people who are different and yet not permitted the individuality of their own unique, non-stereotypical characteristics." Moreover, in the presence of tokens, members of the dominant group seek to clarify for themselves and for others the common characteristics that establish their status and distinguish them from the minority (210–211). Women mathematicians were such a tiny minority of the mathematical community in the forties and fifties that they meet Kanter's criteria for token status; as such they were, indeed, both highly visible and easily stereotyped by many (though not all) of their male colleagues. The postwar intensification of the myth can thus be understood as serving a critical ideological function: to consolidate men's dominance in the mathematical profession.

In her recent book, *Women in Mathematics: The Addition of Difference,* Claudia Henrion (1997) describes in detail the complicated effect that the mathematical community's strongly gendered ideologies have had on recent (especially post-1960) generations of women mathematicians. The women mathematics Ph.D.'s of the forties and fifties, however, were effectively caught in a transition between two cultures. On the one hand, they had many of the women mathematicians of the prewar generation as role models, mentors, and teachers. At the same time, they came to maturity just as the myth of the mathematical life course was gathering momentum.

Faced with a combination of structural discrimination, conflicting cultural messages, and token status, it was indeed difficult for the women mathematicians of the postwar generation to reconcile their mathematical ambitions with the cultural expectations placed upon them. Mary Ellen Rudin, who earned her Ph.D. at the University of Texas in 1949, has characterized her generation of women in mathematics as "the housewives' generation" because they tried at once to meet the cultural expectations of them as women while at the same time maintaining a vital connection to the world of mathematics (Albers, Alexanderson, and Reid 1990: 302). But the reality of the lives and careers of the women mathematics Ph.D.'s of the forties and fifties is much richer and more complex than Rudin's characterization would indicate.

The women mathematicians of the postwar generation were compelled to *improvise:* to create lives and careers that were faithful to their values, talents, and needs by adapting to shifting circumstances and opportunities (see Bateson 1990). Their stories are instructive precisely *because* they do not conform to a set pattern. In the absence of a single, guiding model for how a woman's mathematical life is to be lived, the women mathematics Ph.D.'s of the forties and fifties followed a diversity of paths in the struggle to create a professional identity in post–World War II America.

In the 1990s, women earned mathematics Ph.D.'s at U.S. universities in record numbers. By the mid-1990s, women were receiving approximately one-quarter of the Ph.D.'s awarded each year. If these trends continue, women will rapidly leave behind the token status they have occupied in the mathematical community for over half a century. If women are to have any hope of changing the culture or shaping the direction of the mathematical community of the future, they will need to understand the struggles of the women who have come before them and to rely on the very improvisational skills that these women brought to the process of becoming mathematicians in the past.

In the pages that follow I examine more deeply the social and cultural background, the lives and careers, of the approximately two hundred women who earned Ph.D.'s in mathematics from American colleges and universities during the years 1940 to 1959. In the first chapter, I discuss the role of women in the development of the American mathematical community and the profound transformation of American mathematics wrought by World War II. The myth of the mathematical life course is fully elaborated, and its emergence in the context of the World War II transition is critically examined in detail.

Chapter 2 provides an introduction to the women mathematics Ph.D.'s of the forties and fifties: their background, education, and accomplishments. Special attention is given to a select group of thirty-six women who agreed to participate in intensive oral history interviewing. The remaining chapters are based on the lives and experiences of the thirty-six interviewees, from childhood through retirement and beyond, with a view to understanding how women become mathematicians in the leanest times, when social and cultural forces are least supportive of their ambitions. It is in these times that women rely most heavily on their own resources and inventiveness to persevere in work they love and care about.

Acknowledgments

This project has been several years in the making and would not have come to fruition without the help and support of numerous individuals and organizations. I am indebted to Roberta Mura of Université Laval, whose sabbatical visit to Virginia Tech during the fall semester of 1990 stimulated my thinking about the social history of women in mathematics. I am deeply grateful to Jeanne LaDuke and Judy Green, whose pioneering work on the American women who earned Ph.D.'s in mathematics prior to 1940 has been an inspiration and a resource for my own. In particular, Jeanne LaDuke has been a continual source of support, encouragement, and information for many years; it was she who first suggested to me that the post-1940 Ph.D.'s would be an especially fruitful subject of study.

I am also profoundly indebted to two Virginia Tech colleagues, Susan Hagen and Doris Zallen, who prompted and prodded me to learn more about the theory and practice of oral history, and to Pamela Henson of the Smithsonian Institution, who provided me with the resources and information I needed to get the oral history interviewing underway.

As early as 1994, Thomas Brobson was attentive to my plans for this project and helped immeasurably in my quest for funding. Leslie Kay introduced me to Jean Walton and has maintained a steady interest in my work over several years; her extensive feedback on the manuscript has been invaluable. Rita Kranidis helped me to make crucial decisions about the structure of the book and offered careful readings of interview transcripts as well as early drafts of the first few chapters.

This project was made possible in part by grants from the Spencer Foundation of Chicago, the Alfred P. Sloan Foundation, and the Center for Programs in the Humanities at Virginia Polytechnic Institute and State University, for which I express my deepest gratitude. The statements made and the views expressed in this book are, of course, entirely my own responsibility. Dr. Ted Greenwood of the Sloan Foundation, as well as anonymous reviewers at the Spencer Foundation and MIT Press, provided thoughtful and timely criticisms that greatly improved the final manuscript. Thanks go as well to Amy Pierce Brand of MIT Press, without whom this book might never have come to be.

A preliminary report on this research, based on twenty-three of the thirty-six interviews, was presented at the conference on Women Succeeding in the Sciences, held at Sweet Briar College in April 1997. It will appear in the proceedings of the conference, to be published in 2000 by Purdue University Press. I am grateful to Professor Jody Bart of Sweet Briar for her early interest in my work.

Many other colleagues, friends, and relatives have provided assistance, support, encouragement, and feedback at various stages along the way. Among these, my greatest personal debt is owed to Carol McManus. In addition, I offer heartfelt thanks to Susan Anderson, Moira Baker, Ezra Brown, Laura Clark, Peggy DeWolf, Caren Diefenderfer, Peter Duren,

Thomas Ewing, Deanna Haunsperger, Michael Hughes, Joan Hutchinson, Stephen Kennedy, Cathy Kessel, Jennifer Livesay, Harriet Lord, Laura Mayer, Elizabeth McManus, Roxe Murray, Terry Murray, Sylvia Nasar, Amy Nelson, Kenneth Pimple, Lynn Reed, Barbara Reeves, Margaret Rossiter, Sanford Segal, Sally Sevcik, Daryl Smith, Anita Solow, and Steve Weiss.

Finally, this project would not have been possible at all without the generous participation of the interviewees themselves. I am deeply honored by their willingness to share their lives and stories with me.

A Note on Notation

At various points throughout the text I have found it useful to briefly indicate both the institution and the year in which a specific individual earned her (or his) Ph.D. In such cases I have enclosed the name of the Ph.D.-granting school and the year in which the degree was awarded in parentheses immediately following the name—for example, Mary Ellen Rudin (Ph.D., Texas, 1949). In specific cases I have also indicated the individual's dates of birth and death in the same set of parentheses—for example, Anna Pell Wheeler (Ph.D., Chicago, 1910; 1883–1966).

Citations of oral history interviews are made by giving the last name of the interviewee followed by the page number of the transcript from which the citation is drawn. Two of the interviewees had multiple interviews (and hence multiple transcripts); in these cases, the last name of the interviewee is followed by the transcript number—Anderson-2, Freitag-3. An alphabetical list of the interviewees, which includes the date and location of each interview and the length of each transcript, is given in appendix B.

Illustrations and Tables

Anne Lewis (Anderson)

Winifred Asprey

Lida Barrett

Grace Bates

Barbara Beechler

Jane Cronin Scanlon

Patricia Eberlein

Herta Freitag

Evelyn Granville

Susan Hahn

Anneli Lax

Edith Luchins

Margaret Martin

Cathleen Morawetz

Vivienne Morley

Mary Ellen Rudin

Alice Schafer

Augusta Schurrer

Domina Spencer

Maria Steinberg

Rebekka Struik

Jean Walton

Tilla Weinstein

Tables

Women Becoming Mathematicians

1

Women Mathematicians and the World War II Transition

Men, Women, and the Mathematics Ph.D. prior to 1940

Until very recently, women have been largely absent from accounts of the social and cultural history of mathematics. Prior to 1870, women participated in mathematics, as in the other sciences, as teachers and amateurs but only rarely as professionals.[1] Among the earliest women to achieve recognition for mathematical achievements—Hypatia of Alexandria, Maria Agnesi, Emilie du Chatelet, Sophie Germain, Mary Somerville, and Lady Ada Lovelace—only Hypatia (355?–415) was widely recognized as a bona fide member of the mathematical community of her day; the others were relatively well-to-do European women of the eighteenth and nineteenth centuries who pursued mathematical interests at their leisure (Deakin 1994; Dzielska 1995; Osen 1974; Grinstein and Campbell 1987; Morrow and Perl 1998).

In the mid- to late nineteenth century, however, institutions of higher education began to open their doors to women—in Europe, and, on a grand scale, in the United States—and women began to lay claim to a place in the professions. The doctor of philosophy (Ph.D.) degree was established in Germany in the late eighteenth century, to recognize advanced study and original scholarship in a given discipline. German universities became centers of advanced study and research that were emulated in the United States and around the world. Sonya Kovalevskaya was the first woman to receive a Ph.D. in mathematics, awarded to her *in absentia* by the University of Göttingen in Germany in 1874; she was the first modern woman to pursue mathematics as a profession.[2] Many other women followed as the Ph.D. increasingly came to be acknowledged as the professional certification for research and teaching in advanced mathematics.

But German universities conferred degrees on women only reluctantly; it was in America that women finally broke down the educational barriers and began to earn doctorates in steadily growing numbers. While there had been colleges and universities in the American colonies from the seventeenth century onward—largely devoted to the preparation of young men for careers in the ministry and in law—the development of higher education in the United States was essentially a nineteenth-century phenomenon (Newcomer 1959: 5–34). The rapid industrialization and urbanization of the new nation provided the impetus for this development (Horowitz 1987: 4–7). At midcentury, older institutions such as Harvard and Yale began to refashion themselves as research institutions on the German model (Simpson 1983: 17–21); the Morrill Land-Grant Act of 1862 led to the founding of state colleges and universities in the public interest; and private colleges and universities of every type and description sprouted up across the land in the service of varied educational, vocational, and professional objectives (Rainsford 1972; Wallenstein 1997).

With the proliferation of colleges and universities in nineteenth-century America, higher education became increasingly available to women. Oberlin College in Ohio, founded in 1833, was open to men and women of all races from its inception and awarded the first bachelor's degrees to women in the United States in 1841 (Newcomer 1959: 5; Solomon 1985: 21–22). Women's colleges were established all over the country, particularly in the aftermath of the Civil War; Mount Holyoke, Vassar, Wellesley, Smith, and Bryn Mawr rapidly grew into colleges with outstanding national reputations (Solomon 1985: 47–49). Offered as a compromise between coeducation and sex segregation, the so-called coordinate colleges were usually private women's colleges affiliated with existing public and private men's schools; Barnard, associated with Columbia and founded in 1889, and Radcliffe, associated with Harvard and founded in 1894, are the best-known examples (Newcomer 1959: 40–45; Solomon 1985: 47, 55–56). The state colleges and universities established with Morrill Land-Grant funds were generally coeducational in the North but single-sex in the South (Newcomer 1959; Solomon 1985; Wallenstein 1997).

During the academic year 1861–62, Yale University awarded the first American Ph.D. degrees, and the first such degree in mathematics. In the decades that followed, many other universities developed graduate departments in mathematics and began awarding the Ph.D., most notably Johns Hopkins and Clark Universities and the University of Chicago (Richardson 1989: 365–366; Parshall and Rowe 1994). As research-oriented departments of mathematics emerged in the United States, a genuine *community* of mathematicians began to coalesce. The New York Mathematical Society, founded in 1888, was rechristened six years later as the American Mathematical Society (AMS) to acknowledge its national scope. The AMS, dedicated to the concerns of the growing mathematical research community, sponsored numerous meetings and conferences and published two journals of its own: the *Bulletin* and the *Transactions of the American Mathematical Society* (Parshall and Rowe 1994).

Christine Ladd (later, Ladd-Franklin) was the first woman to be admitted to graduate study in mathematics in the United States, when she began work toward the Ph.D. at Johns Hopkins University in 1878. When she completed her dissertation in 1882, however, Johns Hopkins refused to award her the degree, on the ground that they did not confer doctorates on women; only years later did they relent, awarding her in 1926 the Ph.D. she had earned nearly half a century earlier. Winifred Edgerton was the first woman actually to receive the Ph.D. in mathematics from an American institution, when Columbia University awarded it to her in 1886 (Green and LaDuke 1987, 1989; Grinstein and Campbell 1987).

The struggles of these pioneers opened the mathematical profession to women in the United States. During the nineteenth century, a total of ten American women earned Ph.D.'s in mathematics, nine from institutions in the United States (Bryn Mawr, Columbia, Cornell, and Yale) and one in Europe (at Göttingen). All ten—together with British-born Charlotte Angas Scott, who earned a Sc.D. at the University of London in 1885 and subsequently headed the mathematics department of the newly founded Bryn Mawr College—were active members of the American mathematical community (Green and LaDuke 1987, 1989; Fenster and Parshall 1994).

From 1862 through the first two decades of the twentieth century, the number of mathematics doctorates awarded in the United States

Table 1.1
Ph.D. Degrees in Mathematics Awarded by American Colleges and Universities, 1862 to 1919

Years	Total Awarded	Men		Women	
		Number	Percent	Number	Percent
1862–1869	3	3	100	0	0
1870–1879	10	10	100	0	0
1880–1889	32	31	96.9	1	3.1
1890–1899	84	76	90.5	8	9.5
1900–1909	155	138	89.0	17	11.0
1910–1919	251	216	86.1	35	13.9
1862–1919	**535**	**474**	**88.6**	**61**	**11.4**

Source: Richardson 1989: table 1, 366 (total Ph.D.'s through 1919); Green and LaDuke 1987: table 3, 18 (Ph.D.'s to women through 1919).

increased steadily every year. From 1890 through 1919, the number of these Ph.D.'s awarded to women "doubled each decade" (Green and LaDuke 1987: 15). By 1920, the year American women won the right to vote, sixty-one women had earned the Ph.D. in mathematics from U.S. institutions; women's share of the mathematics Ph.D.'s approached 14 percent of the total. Table 1.1 gives data on the production of mathematics Ph.D.'s in the United States during the years 1862 to 1919, by gender.

Both men and women continued to earn increasing numbers of Ph.D.'s in mathematics in the United States during the 1920s and early 1930s. For both men and women, the numbers of Ph.D.'s earned in the thirties were roughly double those awarded in the twenties (Green and LaDuke 1987: 15–16). During the years 1935 to 1939, there was a drop in the overall numbers of Ph.D.'s awarded, probably due to the Depression. Table 1.2 enumerates the mathematics Ph.D.'s awarded, by gender, during the years 1920 to 1994, reported in five-year aggregates.

Taken together, tables 1.1 and 1.2 reveal some interesting trends. During the teens, twenties, and thirties, women consistently received roughly 14 percent of the Ph.D.'s awarded in mathematics in the United States in each decade. These percentages, while clearly not proportional to the distribution of women in the general population, represent a fairly stable,

Table 1.2
Mathematics Ph.D.'s Awarded in the United States, by Gender (in five-year aggregates, 1920 to 1994)

Years	Total Awarded	Men		Women	
		Number	Percent	Number	Percent
1920–1924	114	93	81.6	21	18.4
1925–1929	238	210	88.2	28	11.8
1930–1934	396	334	84.3	62	15.7
1935–1939	384	333	86.7	51	13.3
1920–1939	**1,132**	**970**	**85.7**	**162**	**14.3**
1940–1944	364	321	88.2	43	11.8
1945–1949	471	427	90.7	44	9.3
1950–1954	1,059	1,008	95.2	51	4.8
1955–1959	1,266	1,208	95.4	58	4.6
1940–1959	**3,160**	**2,964**	**93.8**	**196**	**6.2**
1960–1964	2,082	1,967	94.5	115	5.5
1965–1969	4,325	4,077	94.3	248	5.7
1970–1974	6,187	5,684	91.9	503	8.1
1975–1979	4,690	4,107	87.6	583	12.4
1960–1979	**17,284**	**15,835**	**91.6**	**1,449**	**8.4**
1980–1984	3,591	3,060	85.2	531	14.8
1985–1989	3,765	3,137	83.3	628	16.7
1990–1994	5,253	4,191	79.8	1,062	20.2
1980–1994	**12,609**	**10,388**	**82.4**	**2,221**	**17.6**
1920–1994	**34,185**	**30,157**	**88.2**	**4,028**	**11.8**

Source: Harmon and Soldz 1978, table 26 (1920–1958); National Research Council, Survey of Earned Doctorates 1996 (1958–1994).

healthy share of the Ph.D. total over the period. In particular, by the mid-1930s, women were on the verge of attaining "critical mass": the concentration at which women begin to make the transition from token status to full membership in a professional community.[3] But at that crucial juncture, the momentum of women's gains in mathematics was slowed, first by the Depression and then by the onset of World War II.

About 1935, women's share of the Ph.D.'s in mathematics began a truly startling decline. This decline was at its steepest during the years 1945 to 1955, just as mathematics entered a period of unforeseen power, prestige, and prosperity. From 1945 through 1974, women consistently received fewer than 10 percent of the Ph.D.'s awarded annually. In particular, women's share of the total declined through the entire period 1940 to 1959, reaching an historic low of 4.6 percent in the second half of the 1950s. It was not until the late 1970s that women once again consistently received more than 10 percent of the mathematics Ph.D.'s and not until the 1980s that the percentages returned to the levels attained in the 1930s (Green and LaDuke 1987:12).

The half-century from 1886 to 1936 was an especially bright and promising one for women in the mathematical community. What factors can account for the dramatic decline that came so abruptly afterward? I begin with a look at the American mathematical and social landscape prior to 1940, with particular emphasis on the 1930s, for it is here that the seeds of later developments were sown.

The Polarization of Teaching and Research

From the 1870s onward, as the number of Ph.D.-granting departments of mathematics increased, the American mathematical research enterprise grew in size and international reputation. The AMS, through its journals, meetings, and conferences, supported and encouraged the growth and development of research (Parshall and Rowe 1994). From the late nineteenth century through the mid-1930s, however, a substantial majority of American Ph.D. recipients did little or no research beyond the doctoral dissertation (Richardson 1989: 372–373). The primary professional activity, for nearly all the men and women who received these early American Ph.D.'s in mathematics, was teaching.

Unlike the men, however, women Ph.D.'s usually ended up teaching at women's colleges rather than coeducational colleges or universities. Despite heavy teaching loads, several of these women continued to be active in research and scholarship and assumed leadership positions in the mathematical community. Among the most distinguished of these was Anna Pell Wheeler (Ph.D., Chicago, 1910; 1883–1966), who succeeded Charlotte Angas Scott as head of the Bryn Mawr mathematics department in 1924 and directed seven doctoral dissertations there. Scott and Wheeler numbered among the handful of pre-1920 women doctoral recipients who went on to hold major leadership positions in the AMS in the years prior to 1930 (Green and LaDuke 1987, 1989; Grinstein and Campbell 1982, 1987).

From its very beginnings, the American mathematical community has experienced chronic tension along a fault line that runs between teaching and research. Nowhere was this tension more clearly manifest than in the events surrounding the founding of the Mathematical Association of America (MAA) (Cairns 1938: 1–3; Ewing 1994: 3–4). In 1894, the *American Mathematical Monthly,* a journal dedicated in the main to the concerns of postsecondary mathematics teachers and students, began publication. When, under financial hardship, the *Monthly* sought funding from the AMS in 1914, the AMS politely but firmly refused. This action led to the founding of the MAA two years later. In addition to assuming responsibility for the *Monthly,* the MAA came into being to serve the professional needs of the large community of teachers and students of advanced mathematics. While the two societies had a large core of members in common, there were sufficient differences in membership patterns to suggest a genuine division in the community as a whole (Richardson 1989: 369–370).

In fact, the interests and activities of the early American mathematical community were more varied and diverse than a simple bifurcation into teaching and research would suggest. Particularly during the years prior to 1930, even the most influential American mathematicians were not exclusively devoted to research or to teaching but, rather, to a synthesis of the two into *scholarship* that placed mathematics into historical, critical, and cultural contexts. Even the *Bulletin of the American Mathematical Society* was ostensibly devoted to "a historical and critical review of mathematical science."[4]

But more than any other organization, the MAA served as the natural home for the broader interests and concerns of the mathematical community. Its leadership in the years prior to World War II was firmly committed to scholarship and service to mathematics in the wider sense.[5] In an address delivered in September 1931, MAA President John Wesley Young chastised the AMS for its singleminded focus on research to the exclusion of other mathematical activities (Young 1932). "I have . . . attempted to combat the attitude," Young said, "that would make of research a fetish, that proclaims that the only worthy function of a mathematician is research and that other activities are to be looked on with contempt." He went on to exhort the MAA membership to work in such diverse areas as "criticism, evaluation, and interpretation," "history," the writing of "advanced mathematics from the elementary point of view," "popular exposition," and administration (15).

Young's carefully chosen words reveal the gendered structure of mathematics in the early 1930s, drawing a clear distinction between the "men and women" of the MAA and the "research men" of the AMS. Among his chief priorities is that the MAA continue "to attract men and women of ability to our subject" (10). Indeed, the MAA served as professional home for many women mathematicians. Women published expository articles, posed problems, and offered their solutions in the pages of the *Monthly*, which also regularly reported news of their activities in college and university mathematics clubs. Women presented papers at MAA meetings and served in leadership positions, especially in the regional sections of the Association. The MAA was a place where women were acknowledged as mathematicians, teachers, and scholars, and treated with honor and respect.[6]

During the Depression years of the 1930s, the polarization between research and teaching/scholarship intensified. By 1930, the leadership of the American mathematical community had passed to a younger generation, whose leaders—influenced by spectacular recent developments in atomic chemistry and physics—were far more enamored of narrowly focused research than broad and humanistic scholarship (Parshall and Rowe 1994: 445–451). Stronger ties between industry and basic science, forged in the aftermath of World War I, had already begun to favor mathematical research in the 1920s. Industrial philanthropies like the Rockefeller and Carnegie Foundations supported research in mathematics and science both directly and through the National Research Council (Green and LaDuke 1987: 16; Kleinman 1995: 31–45).

While many creative young mathematicians were already present, the resurgence of American mathematical research was fueled by the infusion of mathematical immigrants from central and eastern Europe that began in the 1920s and gained momentum with Hitler's rise to power in the 1930s (Dresden 1942; Bers 1989; Reingold 1989). Many of these immigrants had experienced the heady intellectual atmosphere of the major European research centers in Vienna, Göttingen, and elsewhere. There was an eagerness in the American mathematical research community to recreate this exciting atmosphere in the United States, and several émigré mathematicians had a hand in bringing these dreams to fruition.

Among the most significant developments of the prewar period was the founding of the Institute for Advanced Study at Princeton in 1930. The Institute, financed by corporate philanthropy, was conceived as a center for postgraduate research, with a small permanent faculty and a large cadre of temporary members holding short- and long-term appointments. The School of Mathematics was the first department created in the Institute, its original faculty a mix of Americans—James Alexander, Marston Morse, and Oswald Veblen—and refugees—Albert Einstein, John von Neumann, and Hermann Weyl.[7]

While the thirties brought renewed enthusiasm (and financial support) for research to American mathematics, the excitement was rather unevenly distributed, even among the most prestigious graduate departments. At the University of Chicago, for example, the faculty had become somewhat inbred. Numerous Chicago Ph.D.'s had simply joined the faculty on graduation, and the research emphasis of the department lay in directions that had been more fashionable several decades earlier (MacLane 1989: 138–143; Parshall and Rowe 1994: 445–446). Although many distinguished mathematicians came to the United States as refugees in the 1930s, they were not immediately made welcome. Many were unable to find employment commensurate with their talents in research due to anti-Semitism, particularly in the form of Jewish faculty quotas at several prestigious institutions (Bers 1989; Reingold 1989). Moreover, the financial constraints of the Depression often made it impossible for departments to engage in much recruitment of new faculty or to create conditions conducive to research productivity (Niven 1989).

At the same time, despite the heightened visibility of mathematical research in the thirties, the broadly based, humanistic scholarship of the early years did not die. While the *Bulletin of the AMS* quietly dropped its historical and critical missions in 1931, new journals devoted to historical and philosophical aspects of mathematics, such as *Scripta*

Mathematica, appeared on the scene (Merzbach 1989: 654). George Sarton, the highly influential Harvard historian of science, argued passionately in 1936 for a continuation of the thorough, careful study of the history of mathematics—by mathematicians (Sarton 1936).

But in terms of prestige and power, mathematical research was clearly in ascendancy. While the MAA and its functions, as laid out so eloquently by John Wesley Young, continued to survive and to thrive, it was clear by the late thirties that power and influence were shifting to "the research men." Although distinguished and highly productive women mathematicians came to the United States in the European migration—Emmy Noether and Olga Taussky most notable among them—the growing prestige of research made it increasingly the province of men. It is not surprising, then, that while women continued to hold national and regional leadership positions in the MAA after 1930, they were virtually absent from leadership positions in the AMS from the 1930s through the 1960s (Green and LaDuke 1989: 386, 390).

In fact, the Depression was a time of setbacks for women after the tremendous educational, social, political, and professional gains they had made through the twenties (Brown 1987; Ware 1982). Working women, both in academia and elsewhere, were particularly singled out for pay cuts, demotions, and dismissals, justified by the argument that men raising families should be accorded preference in a time of diminished resources (Rossiter 1995; Ware 1982). The cultural and economic sea change of the thirties offered but a glimpse of the greater turbulence to come.

Women and the "Mathematician's War"

Before World War II, virtually no government grants were available for mathematical research. Apart from agricultural research at universities and research conducted in federal laboratories, the government sponsored comparatively little scientific research of any kind. Indeed, the only grants available for basic science at colleges and universities were funded by private foundations. Large corporations carried out their own product-related research, but very little of this was of a mathematical nature (Kleinman 1995: 26–50).

But as war broke out and spread across Europe, the U.S. government came to recognize the importance of science and technology to national security. In June 1940, well over a year before the American entrance into World War II, President Roosevelt created the National Defense Research Council (NDRC), which was incorporated a year later into the Office of Scientific Research and Development (OSRD) under the directorship of Vannevar Bush. The OSRD funded and coordinated war-related research and development efforts through numerous university and industrial contracts (Kleinman 1995: 61–65; Zachary 1997).

Mathematicians played key roles in many of these efforts, largely under the auspices of the Applied Mathematics Panel of OSRD, directed by former Wisconsin mathematics department head Warren Weaver (Rees 1989). So great was the involvement of mathematicians in the war effort that Frank B. Jewett, chair of Bell Telephone Laboratories and an adviser to Vannevar Bush, went so far as to characterize World War II as "a mathematician's war" (Jewett 1942: 240).

What role did women play in this "mathematician's war"? Very few women mathematicians are mentioned in contemporary and retrospective accounts of war-related research and development. In fact, the contributions of just two pre-1940 Ph.D.'s recur in later accounts of mathematicians' wartime activities. Mina Rees (Ph.D., Chicago, 1931) served as technical aide to Warren Weaver on the Applied Mathematics Panel and subsequently played an important role in postwar government funding of mathematical research. Grace Murray Hopper (Ph.D., Yale, 1934) joined the Naval Reserve and began an involvement with the navy and the development of high-level computer languages that lasted for well over forty years.[8]

The role played by women mathematicians on college and university campuses was more prominent during wartime, however. As young men disappeared into the armed forces and faculty members disappeared into war work, both undergraduate and graduate enrollments on American college and university campuses were down, sometimes dramatically (Olson 1974: 46). In the wartime environment, women were especially welcome as students to offset declining enrollments. At Virginia Polytechnic Institute, a nominally coeducational but predominantly male land-grant institution, the historian Peter Wallenstein writes, "the pressure to include women grew stronger, as students, especially men, grew scarce. . . . [T]he number of regular students dropped from 3,382 in spring 1942 . . . [to] 557 in spring 1944, to only 411 in spring 1945" (1997: 155 and 161).

In all of the sciences, and mathematics in particular, women were also made welcome as graduate teaching assistants, to offset staffing shortages caused by involvement of male faculty in off-campus war research. As the war progressed, the government instituted special instructional programs for military personnel that were to be carried out under contract on college and university campuses, most notably the Navy V-12 program and the Army Specialized Training Program (ASTP). At the approximately 300 colleges and universities sponsoring the V-12 and ASTP programs, the demand for teaching personnel escalated. Students and recent Ph.D. graduates in mathematics, male and female, were often called on to staff these military courses (Newsom 1943a, 1943b; Cairns, Dresden, and Kline 1943; Price 1943).[9]

The wartime enrollment decline was decisively reversed with the passage of the Servicemen's Readjustment Act of 1944, familiarly known as the G.I. Bill of Rights. Even before the war's end, veterans who had seen a tour of military duty began enrolling in college and university courses, often for the first time. The demand for teaching personnel, which had gone up slightly and selectively with the institution of the military training programs, increased substantially across the board beginning in 1944 (Olson 1974). From 1942 to 1946, the number of women teaching mathematics at U.S. colleges and universities more than doubled, while the number of women on faculties across all the science disciplines more than tripled. Indeed, as historian Margaret Rossiter asserts, "the movement of women scientists onto college and university faculties during the war was so widespread that it constituted their principal contribution to the war effort" (1995: 10–11).

In short, while the crisis of World War II brought increased opportunities for women to study and teach advanced mathematics on college and university campuses, it also tended to heighten the polarization of teaching and research in mathematics that had begun in the 1930s. While "research men" had helped to win the war, women took up the task of educating the troops.

Mathematics, Education, and National Security in the Postwar Years

Because the Allies' victory in World War II was seen as having resulted significantly from their scientific and technological superiority, political sentiment ran strongly in favor of continued governmental and military support of basic research. The postwar years saw the establishment in 1946 of the Office of Naval Research (ONR) and the Atomic Energy Commission (AEC), and finally, in 1950, the National Science Foundation (NSF). Because so many mathematicians had been involved in war work, new government funding agencies were extremely generous to projects in all areas of mathematics, pure and applied. This most basic, least empirical, and "purest" of the sciences was now seen as crucial to the security and power of the nation (Richardson 1943; Brink 1944; Rees 1989; Kleinman 1995; Zachary 1997).

After the austerity of the Depression and World War II, American mathematics was poised on the brink of a period of growth and efflorescence. All that was needed to fuel the expansion was a massive influx of young recruits to the field. That influx was provided, at least initially, by the G.I. Bill, which enabled millions of American service personnel—some of them women, but the overwhelming majority of them men—to undertake undergraduate and graduate study in colleges and universities all across the United States. Many of these men would not have attended college at all without the financial assistance of the G.I. Bill, and many of them, on graduation from college, found academic life to their liking and continued on for advanced degrees (Olson 1974; Hartmann 1982).

The glut of incoming students created a dramatic shortage of teachers; graduate students and even knowledgeable undergraduates were pressed into classroom service.[10] In the early years after the war, with graduate enrollments still comparatively low, women were generally made to feel welcome as graduate students in mathematics; their services as teachers were particularly in demand. At the same time, many major public and private universities coped with the enrollment boom by restricting undergraduate admissions of women. Even some women's colleges made provisions to accommodate returning G.I.'s on their campuses.[11]

By the late 1940s, increasing numbers of men were starting work toward the Ph.D. in mathematics. Women, who had been welcomed as graduate students throughout the war and amidst the chaos of the early postwar years, suddenly found themselves in competition with substantial numbers of men for faculty attention and departmental resources. As the environment for graduate students became more competitive, so, too, did the environment for faculty, both on university campuses and in the mathematical community as a whole. With the increased availability of research funds from the federal government, and graduate teaching assistants to do lower-division teaching, graduate departments of mathematics had the financial means to lower the teaching loads of their faculty involved in research. At the same time, newly-established federal grants provided summer salaries and release time from teaching. With increased means came increased expectations of faculty research productivity; ambitious younger faculty could choose, from a large pool of talent, those doctoral students most likely to assist them in carrying out their research plans.

A host of factors contributed to the declining share of mathematics Ph.D.'s awarded to women in the 1950s. Because of the stateside shortage of men during World War II, women were allowed and even encouraged to work in industries and professions that were traditionally closed to them. The image of Rosie the Riveter on patriotic posters of the time enhanced the notion that working women were active contributors to the national defense. But once the war had been won, it was time for women to return to their traditional roles. The prevailing cultural message to the woman of the late forties and early fifties—delivered through magazines, newspapers, radio, and eventually television—was that her proper social role was to marry, make a home, bear and raise children, and provide emotional nurturance and social support to her husband. To do so was just as patriotic as working outside the home during the war had been (Friedan 1963; Hartmann 1982).[12]

Just as young men were increasingly drawn to academic careers in mathematics, academic institutions were erecting barriers to women's professional advancement. Antinepotism rules became increasingly prevalent at American colleges and universities and came in many forms. Some prohibited members of the same family from being employed in the same institution or the same department, and still others prohibited one family member from being in the position to make personnel decisions about another. What they all had in common was that, in practice, they were applied prejudicially and often peculiarly against women (Dolan and Davis 1960; Rossiter 1982, 1995). Because women mathematicians who marry frequently marry other mathematicians, antinepotism laws could be particularly crippling to the career prospects of a married woman in mathematics.[13]

The expansion in higher education that had begun in response to the G.I. Bill continued through the 1950s and into the 1960s. Once the United States had entered the cold war, the highest priority was placed on maintaining the technological and military superiority of the United States over the Soviet Union and its allies. The Soviet launching of the first space satellite, *Sputnik,* in 1957, merely intensified the preexisting emphasis on scientific research and development and, concurrently, on scientific and technical education (Clowse 1981; Rossiter 1995: 63–64).

The continued growth of higher education was ensured by the demographic pressure of the baby-boom generation—the children born between 1946 and 1964—which brought a new wave of bulging undergraduate enrollments in the sixties. As the enterprise of American higher education expanded, colleges and universities continually sought to upgrade their facilities and enhance their status. "Upgrading and enhancement" frequently meant heightened emphasis on faculty research and graduate education. In particular, state normal schools and teachers colleges were transformed into comprehensive state universities. Women's colleges felt increasing pressure to become coeducational and to increase the proportion of men on their faculties (Harcleroad and Ostar 1987: 41–64; Rossiter 1995: 206–234). The net effect of these transformations was to substantially reduce the visibility of women in academia.

The Myth of the Mathematical Life Course

The twenty-five-year period following the conclusion of World War II was, in the opinion of many observers, the golden age of American mathematical research. As the prospect of a career in mathematics became increasingly attractive and desirable in the prosperous postwar years, *the myth of the mathematical life course* (hereinafter, *the myth*) became the prevailing model of how that career should unfold. While many aspects of the myth date back to the nineteenth century and earlier, the conditions were ripe in the postwar years of the forties and fifties for the myth of the mathematical life course to grow and flourish. The rapid expansion of higher education, the proliferation of academic job openings, the high status and substantial funding accorded to mathematical research: all of these circumstances created the conditions under which it was possible, for the first and perhaps only time in history, for substantial numbers of mathematicians—mostly men—to carry on their lives and careers in conformity with its pattern.

According to the myth, mathematical talent and creative potential emerge very early in childhood. These natural gifts are focused and directed toward mathematics from this early age. In college, that the major will be mathematics is a foregone conclusion, and the student proceeds from college to an elite graduate school without a break. In graduate school, the student comes under the tutelage of a powerful mentor, under whose direction he writes a doctoral dissertation that makes a significant contribution to his area of study. As his graduate studies draw to a close, his mentor assists him in landing a postdoctoral research position at a similarly elite doctorate-granting department of mathematics, and afterward he goes on to one or more positions at comparably distinguished universities, where his creative achievements are rewarded with tenure.

The mathematician is extraordinarily productive in research from his late teens until his early forties and during this period does his best work. There are no interruptions during these years of scholarly productivity (except possibly for a brief period of military service), and to a considerable extent, the mathematician ignores or eschews other interests during this time. It is very helpful if the mathematician has a spouse who will take care of domestic and family concerns and provide him with a peaceful home environment that supports his creative work.

In the later years, research productivity continues, albeit at a somewhat lesser rate; the mathematician continues to generate creative ideas, but the working-out of these ideas falls to his younger colleagues and (especially) graduate students, who carry out the various aspects of his research program. It is perhaps possible, later in life, for the mathematician to enjoy some hobbies and diversions, but his primary concern is and continues to be mathematics.

Several aspects of the myth are clearly peculiar to mathematics. Because many people have experienced school mathematics as abstract and difficult, the general public is inclined to view mathematics as the domain of geniuses who understand and enjoy the subject because they were "born that way." Since mathematical knowledge is cumulative, early difficulty with mathematics often seems to be compounded with the passage of time; so that by the sixth grade, for example, it may already be too late to repair a wounded relationship with the subject. Such experiences seem to underscore the popular perception that mathematical talent needs to be discovered and nurtured from an early age (Tobias 1993).

In many important respects, however, the myth is not specific to mathematics but describes the ideal development of any academic career. In an incisive analysis of the academic career system as it has evolved over the past two centuries, the sociologist Arlie Russell Hochschild (1975) has pointed out that many disciplines emphasize the importance of creative achievement at an early age. The emphasis on uninterrupted progress is not even unique to academic life but underlies the whole notion of how careers—especially men's careers—ought to unfold. Her characterization of the idealized American academic career path sounds strikingly similar to the timetable laid out in the myth. Moreover, she explicitly questions the appropriateness of this timetable for women:

The academic career is founded on some peculiar assumptions about the relation between doing work and competing with others, competing with others and getting credit for work, getting credit and building a reputation, building a reputation and doing it while you're young, doing it while you're young and hoarding scarce time, hoarding scarce time and minimizing family life, minimizing family life and leaving it to your wife—the chain of experiences that seems to anchor the traditional academic career. Even if the meritocracy worked perfectly . . . I suspect there would remain in a system that defines careers this way only a handful of women at the top (49).

The emphasis on youthful achievement, common to the folklore about both mathematicians and academics, seems to be significantly a product of the modern conception of the career. "The link between age and achievement for many specialties housed in the university resembles that of athletes more than that of popes or judges," Hochschild writes. "Interestingly, achievement came later in life for men before 1775—before the massive bureaucratization of work into the career system" (61).

How many mathematicians, of either the prewar or the postwar period, have managed to live out the letter of the myth?[14] Regardless of the extent to which the myth is mirrored in the reality of actual lives, there can be little doubt of its power. The myth has been, and continues to be, aggressively perpetrated, particularly in interviews, autobiographies, biographies, memoirs, and reminiscences of mathematicians whose careers reached their full flowering during the postwar expansion.[15]

In a number of important particulars, the myth is an inappropriate one for women, and for this reason I have used the masculine pronoun throughout my description of its major features. Few women have spouses who will shield them from distractions or exempt them from the responsibilities of home and family. Those women who have such supportive spouses nevertheless must face the essential conflict between single-minded focus on work and the biological timetables of reproduction. In fact, it is only relatively recently that women have been able to combine marriage and academic career at all: before World War II, female elementary and secondary schoolteachers were routinely dismissed on marriage, and women faculty at women's colleges were expected to remain single (Ware 1982; Scharf 1980; Rossiter 1982; Solomon 1985). Historically, even unmarried or childless women have experienced barriers to their educational and career advancement that have made youthful achievement and professional continuity extremely difficult (Ahern and Scott 1981: xviii; Ruskai 1984).

The women mathematicians of the wartime and postwar generations came to maturity just as the myth was gathering strength in the American mathematical community. But their teachers and mentors included men and women of the prewar generation, who held to a broader view of the mathematical community—embracing both women and men—and a more inclusive view of mathematical activity—encompassing research, teaching, and scholarship. The women mathematics Ph.D.'s of the forties and fifties were caught between two cultures. While many attempted to conform to the myth, experiencing conflict in the attempt to do so, some found ways to adapt the myth to their own life circumstances. Others were able to pursue careers on the model of the prewar generation of women in mathematics. Most often, however, their mathematical lives were an improvisational blending of the old and the new.

What did it mean to the women of this transitional generation to be a mathematician? Were they able to develop, nurture, and sustain a sense of mathematical identity, particularly in cases of extreme dissonance between their own aspirations and the contradictions of the mathematical community and the larger culture? In the chapters that follow, I take a closer look at the group of women who received Ph.D.'s in mathematics from American colleges and universities during the years 1940 to 1959. In particular, I consider the various stages of mathematical development, as viewed through the eyes of thirty-six of these women, in an attempt to obtain some insight into the answers to these questions.

2

Women Mathematics Ph.D.'s of the 1940s and 1950s

Finding the Women

During the years 1940 through 1959, approximately two hundred Ph.D.'s in mathematics were awarded to women by colleges and universities in the United States, as enumerated in table 1.2 of chapter 1. Because of inevitable inaccuracies in reporting and collection of data, the precise numbers are somewhat uncertain. Rounding the statistics to the nearest ten, roughly ninety women and 750 men were awarded the Ph.D. degree in mathematics by American institutions during the 1940s, while 110 women and 2,220 men received the degree in the 1950s.

During the second half of the forties and the decade of the fifties, the sheer number of Ph.D.'s in mathematics awarded in the United States increased so dramatically that it is virtually impossible to obtain a complete list of the awardees. Because they constitute a relatively small subset of the total, however, the task of compiling a fairly complete list of the women is a much more manageable one.

My search for the women began with the *Bulletin of the American Mathematical Society* for the years 1941 through 1960. In each May issue during this period, the AMS printed a complete (or nearly so) list of the previous year's Ph.D. recipients in the mathematical sciences, including pure and applied mathematics, mathematical statistics, mathematical physics, and history of mathematics. It was the practice of the AMS, at least until the late 1950s, to print the full names of women and only the initials of men. It was thereby possible to locate, on the basis of names readily identifiable by gender, well over 150 of the female Ph.D.'s.

To locate the rest, I consulted several other sources. In the late 1970s, Amy King and Rosemary McCroskey (1976–1977) compiled a list of all the women mathematicians they could locate in the United States and Canada. While their list is incomplete and contains a number of errors, it provides a starting point for further searches. One of the most vexing problems in trying to identify the women mathematicians of this generation is that women's surnames usually changed on marriage; the King-McCroskey list was particularly helpful in tracking down name changes. Still other women mathematicians were identified by their appearance in standard reference works, particularly *American Men of Science (AMoS)*—later renamed *American Men and Women of Science (AMWoS)*—and *Dissertation Abstracts*.

Once I had compiled a basic list from these sources, I sought direct verification of the names by directly contacting the women themselves— or, failing that, the graduate departments or alumni associations of their graduate institutions. As a result of these exhaustive searches, I was able to compile a more definitive list of 188 women, eighty-three from the 1940s and 105 from the 1950s, who received Ph.D.'s in mathematics from American institutions during the years 1940 to 1959.

While basic biographical information on many of these women can be found in library sources (*AMWoS*, once again, serving as a good starting point), it was my aim to try to speak to as many of them as possible to come to a fuller understanding of their lifelong relationship with mathematics. With this goal, I began searching for their names in the membership directories of the major mathematical and statistical societies. I was able to identify, locate, and contact those women who had remained active mathematically to the extent that they still maintained membership in one or more professional associations or maintained contact with a network of colleagues and friends in academia.

In the end, I conducted in-depth, in-person interviews with thirty-six of the original two hundred women. Appendix A provides a discussion of how the interviewing was carried out. The interview group, while decidedly not a *random* sampling of the women Ph.D.'s of the period, is a *representative* sampling of the professionally active women mathematicians of this generation. Seventeen of them received doctorates in the 1940s, nineteen in the 1950s. Together, they constitute nearly 20 percent of the women who received Ph.D.'s in the mathematical sciences during the two decades.

A Distinguished Cohort of Mathematical Women

The women who earned their Ph.D.'s in mathematics during the forties and fifties include some of the most distinguished mathematicians and mathematics educators of this century. Julia Robinson, Cathleen Morawetz, and Mary Ellen Rudin are probably the best-known women mathematicians of this generation.[1] The logician Julia Bowman Robinson (Ph.D., California/Berkeley, 1948; 1919–1985) did pioneering research leading to the solution of Hilbert's Tenth Problem, a long-standing open problem in the logical foundations of arithmetic. In recognition of her research achievement, she was elected to the mathematics section of the National Academy of Sciences in 1976, the first woman to be so honored. In 1982, she was elected the first woman president of the American Mathematical Society, and in 1983, she was awarded a MacArthur Foundation Fellowship.

Cathleen Morawetz (Ph.D., NYU, 1951), who has had a long and productive career in applied mathematics at NYU and has served as director of the Courant Institute of Mathematical Sciences there, was the first woman elected to the applied mathematics section of the National Academy of Sciences in 1990. She was elected president of the AMS in 1995, the second woman to hold that office, and in 1998 she was awarded the National Medal of Science. Mary Ellen Rudin (Ph.D., Texas, 1949) has done creative work in the field of set-theoretic topology over a span of fifty years; the subject of several popular interviews, she has been an especially energetic and visible figure in the American mathematical community.

Many of the other women Ph.D.'s of the forties and fifties have made research contributions that have equaled and even surpassed these three. Among those who have made particularly substantial contributions are Dorothy Maharam Stone (Ph.D., Bryn Mawr, 1940) and Alexandra Bellow (Ph.D., Yale, 1959) in measure theory and ergodic theory; Domina Eberle Spencer (Ph.D., MIT, 1942) in mathematics applied to physics and engineering; Josephine M. Mitchell (Ph.D., Bryn Mawr, 1942) in classical and complex analysis; Jane Cronin Scanlon (Ph.D., Michigan, 1949) in functional analysis, differential equations, and mathematical biology; Esther Seiden (Ph.D., California/Berkeley, 1949) in mathematical statistics; Wanda Szmielew (Ph.D., California/Berkeley, 1950; 1918–1976) and Marian Boykan Pour-El (Ph.D., Harvard/Radcliffe, 1958) in logic; Mary Bishop Weiss (Ph.D., Chicago, 1957; 1930-1966) in harmonic analysis; Vera Pless (Ph.D., Northwestern, 1957) in coding theory and combinatorics; and Tilla (Klotz Milnor) Weinstein (Ph.D., NYU, 1959) in complex analysis and geometry.

Several women in the group have been especially prolific in mathematics education, among them Mary P. Dolciani (Ph.D., Cornell, 1947; 1923–1985) and Margaret F. Willerding (Ph.D., St. Louis, 1947), who are well-known to generations of students at the college and precollege level for their numerous mathematics textbooks. Many others have devoted part or all of their careers to college and university administration, including two who earned doctorates at the University of Pennsylvania: Jean B. Walton (Ph.D., 1948) was dean of women, and later dean of students and coordinator of women's studies, at Pomona College; Lida K. Barrett (Ph.D., 1954) served as department head, associate provost, and dean at three different universities and in 1989 became the second woman elected president of the MAA.[2]

Quite a few women in the doctoral classes of the forties and fifties have had distinguished careers in the federal government. Notable among these are Leila D. Bram (Ph.D., Pennsylvania, 1951; died 1979), for many years the director of the mathematics program at the Office of Naval Research; Elizabeth H. Cuthill (Ph.D., Minnesota, 1951), who spent her forty-year career at the David W. Taylor Naval Research and Development Center at Annapolis; Ruth M. Davis (Ph.D., Maryland, 1955), who served as Assistant Secretary of Energy in the Carter administration; and Joan R. Rosenblatt (Ph.D., North Carolina, 1956), who had a distinguished career at the National Bureau of Standards (NBS)—later the National Institute for Standards and Technology (NIST)—from which she retired as director of the computing and applied mathematics laboratory in 1996.

At least twenty-three Roman Catholic nuns earned Ph.D.'s in mathematics from 1940 to 1959. Many of these women undertook graduate study comparatively late in life, after they had been teaching at the college level for a number of years. Few of them did much research beyond the Ph.D. However, two of the nuns who earned doctorates at the University of Michigan during the 1940s have achieved some notoriety recently for their research. Sister Mary Claudia Zeller (Ph.D., Michigan, 1944; 1910–1991?) wrote her dissertation on the history of trigonometry. The historian of mathematics Ezra Brown has made use of her work in his study of the medieval mathematician Regiomontanus (Brown 1990). Sister Mary Celine Fasenmyer (Ph.D., Michigan, 1946; 1906–1996) completed her Ph.D. in the field of special functions under the direction of Earl Rainville. She published two papers based on her dissertation and had a long teaching career at Mercyhurst College in Erie, Pennsylvania. Rainville (1960) showcased her doctoral work in a widely read textbook, and she has recently been rediscovered by a new generation of researchers in the area of special functions and combinatorics (Zeilberger 1982; Hutchinson 1994; Warnick 1994).

In a generation of women mathematicians that features a substantial number of firsts, we find the first two African American women to earn Ph.D.'s in mathematics. Evelyn Boyd Granville (Ph.D., Yale, 1949) and Marjorie Lee Browne (Ph.D., Michigan, 1950; 1914–1979) earned their doctorates within just a few months of one another. No African American woman would earn a Ph.D. in mathematics again until the early 1960s. Marjorie Lee Browne earned her Ph.D. at the age of thirty-five and had a thirty-year teaching career at North Carolina Central University. Evelyn Granville pursued her undergraduate and graduate studies in mathematics without a break, earning her Ph.D. at the age of twenty-five; she went on to a career in industry and academia spanning nearly fifty years.[3]

Leading Producers of Women Ph.D.'s: Continuity and Change

Which U.S. institutions graduated the largest numbers of women mathematics Ph.D.'s in the postwar years? Table 2.1 lists the leading producers of women Ph.D.'s in the forties and fifties, by decade; for comparison, the table also includes the leading producers for the decade of the thirties and overall for the period 1886 to 1939. In each case, the institutions listed account for well over half the Ph.D.'s in mathematics awarded to women in the given period.

Table 2.1
Ranking of Institutions by Number of Mathematics Ph.D.'s Awarded to Women

1886–1939 (10 or more)		1930–1939 (4 or more)	
46	Chicago	24	Chicago
21	Cornell	13	Catholic
19	Bryn Mawr	8	Bryn Mawr
15	Catholic	8	Cornell
13	Yale	8	Illinois
12	Johns Hopkins	6	Radcliffe
12	Illinois	67	Six schools
138	Seven schools		

1940–1949 (4 or more)		1950–1959 (4 or more)	
9	Illinois	10	NYU
8	Catholic	6	Brown
8	Michigan	5	Illinois
7	Radcliffe	5	Michigan
6	Chicago	5	Minnesota
5	California/Berkeley	4	California/Berkeley
4	Cornell	4	Catholic
47	Seven schools	4	Radcliffe
		4	St. Louis
		47	Nine schools

Source: For pre-1940 data: Green and LaDuke 1987: 18 (table 3).

The table reveals some interesting historical trends. First of all, many of the older graduate departments declined in importance as producers of women Ph.D.'s in mathematics. At Johns Hopkins, Yale, Chicago, and Cornell, departments that had been granting Ph.D.'s in mathematics since the late nineteenth century, the decline may be attributed in part to the aging of the faculty and the subsequent reorganization of individual departments around a new generation of researchers. After reorganization, these departments did not resume their former status as major producers of women Ph.D.'s.

The decline at Chicago was especially dramatic. During the years 1940 to 1946, Chicago conferred six mathematics Ph.D.'s on women; in the thirteen years following, Chicago awarded only three degrees to women (in 1951, 1956, and 1957). As it happens, the year 1946 was a key turning point in the history of the department at Chicago. In that year, Marshall Stone came to Chicago from Harvard to head the mathematics department, and the years immediately following his arrival were marked by a nearly complete turnover in the faculty (MacLane 1989). In particular, the people who had supervised the majority of women's dissertations in mathematics in the preceding years had by that time either retired, died, or departed for other reasons. Among those who retired in 1946 was Mayme Logsdon (Ph.D., Chicago, 1921; 1881–1967), a tenured associate professor. According to Saunders MacLane, "one of her duties was that of advising and helping many women who were graduate students at Chicago" (1989: 135). After her departure, Chicago would not have a tenured woman in the mathematics department for another forty years.[4]

Stone, charged with the responsibility of creating a first-rate research department, hired a number of émigré mathematicians, along with a large cadre of young men at an early stage in their careers. The university, ideally situated in a major urban center, was poised to take full advantage of the surge of graduate students produced by the G.I. Bill. But despite the surging male enrollments, approximately 14 percent of the students enrolled in graduate mathematics courses at Chicago during the late forties and fifties were women—about the same rate at which women earned Ph.D.'s in mathematics prior to 1940 (MacLane 1989: 145–146). Yet few of these women actually completed the degree. Clearly, something in the environment for women at Chicago had changed.

The two women's colleges that had historically granted large numbers of mathematics Ph.D.'s to women exhibit diverging patterns in the forties: Bryn Mawr disappears from the leaders entirely, while Radcliffe continues to appear in the list, albeit with diminishing numbers of graduates. The graduate school at Bryn Mawr became coeducational in the 1930s and began to admit men (Rossiter 1995: 91). Moreover, the tiny Bryn Mawr mathematics department lost its energetic leader, Anna Pell Wheeler, to retirement in 1948; her last student, Dorothy Maharam Stone, received her Ph.D. in 1940 (Grinstein and Campbell 1982). The features that had made graduate study at Bryn Mawr unique since the time of its founding were rapidly disappearing, and women students at

Bryn Mawr were often advised to go elsewhere for doctoral work. By contrast, women who earned Radcliffe doctorates took their coursework and pursued research with advisers on the Harvard faculty. While many women decided against attending Radcliffe because they would be denied the Harvard Ph.D., many others were attracted to graduate study there because of the quality of the Harvard mathematics program.[5]

The Big Ten universities of the Midwest—including Illinois and Michigan and, later, Minnesota—were major educators of women mathematicians in the forties and fifties; Illinois had been among the leaders already in the thirties. The relatively late appearance of these schools among the top producers of women Ph.D.'s is, in part, a reflection of the fact that their graduate programs developed a few decades later than those at the well-known private universities. Michigan, in particular, rapidly developed an outstanding reputation for both teaching and research in the 1930s. Two of its faculty, Raymond L. Wilder and the Russian émigré G.Y. Rainich, were excellent teachers and helped to create an environment for undergraduate and graduate students that was both personally welcoming and intellectually stimulating (Kaplan 1989). Both Wilder and Rainich directed a number of dissertations by women in the forties and fifties. Both Illinois and Michigan—like their prestigious private predecessors, such as Chicago and Cornell—nevertheless show a decline in the number of Ph.D.'s awarded to women from the forties to the fifties.

The University of California at Berkeley appears in the leader list for the first time in the 1940s and remains there in the 1950s. Its presence there is due almost entirely to the influence of two men: the Polish émigrés Jerzy Neyman (1894–1981) and Alfred Tarski (1902–1983), who came to Berkeley in 1938 and 1942, respectively. The statistician Jerzy Neyman supervised the dissertations of two women, Evelyn Fix (Ph.D., Berkeley, 1948; 1904–1965) and Esther Seiden (Ph.D., Berkeley, 1949), and consulted on yet a third, that of Elizabeth Scott (1917–1988), who earned a Ph.D. in astronomy in 1949. Neyman had a reputation for being especially supportive of women colleagues, both at Berkeley and elsewhere. Both Fix and Scott remained on the statistics faculty at Berkeley for the rest of their careers. Scott, in particular, became one of Neyman's most important research collaborators and an eminent statistician in her own right. Tarski supervised the dissertations of four women students in mathematical logic during the years 1948 to 1953; Julia Robinson was among the first of these.[6]

The Catholic University of America was founded in the 1880s for the graduate education of the Catholic priests and brothers who were to teach at Catholic men's colleges. It was only much later that the Catholic women's colleges began to send the nuns on their faculties to graduate school as well (Power 1972; Solomon 1985). As a result, America's premier Catholic graduate school emerged as a major grantor of mathematics Ph.D.'s to women in the 1930s (Green and LaDuke 1987). In the forties and fifties, however, as Ph.D.-granting departments grew in number and became more geographically widespread, Catholic University declined in importance, as many other institutions—both Roman Catholic (such as St. Louis and Notre Dame) and secular—awarded doctorates to nuns.

The emergence of NYU and Brown as the top two producers of women Ph.D.'s in mathematics during the 1950s is dramatic and striking. These two universities share the distinction of having started, during the 1940s, the first graduate programs in applied mathematics in the United States (Rees 1989). At both institutions, women were enthusiastically invited to join the growing ranks of this field. At Brown, six of the seven women who received mathematics Ph.D.'s in the years 1948 to 1958 earned them in applied mathematics.

At NYU in the late thirties and forties, Richard Courant built what came to be known first as the Institute for Mathematics and Mechanics and, eventually, the Courant Institute of Mathematical Sciences. Through a judicious recruitment of European émigrés and young American talent, Courant presided over the rapid development of pure and applied mathematics at NYU. Courant was drawn to bright, talented young women and established a reputation for encouraging them to pursue careers in mathematics.[7] All three Ph.D.'s in mathematics awarded to women at NYU during the 1940s were directed by Courant; and while only one of the ten women Ph.D.'s of the 1950s was supervised by him, his influence was felt by nearly *all* the women who studied at NYU during those years, as interviews with the women bear out. The atmosphere at NYU—particularly during the late forties and early fifties—has been described as intellectually lively, friendly, warm, and familial.

Just as certain graduate departments were noted for the production of comparatively large numbers of women Ph.D.'s, certain advisers were noted for their encouragement and willingness to advise and support women students. At NYU, for example, two pure mathematicians, Wilhelm Magnus and Lipman Bers, came to be known as especially friendly to women; each supervised the dissertations of three women students in the 1950s and many more thereafter.[8] In general, émigré mathematicians were often particularly encouraging to women. To those I have already mentioned—Bers, Courant, Magnus, Neyman, Rainich, and Tarski— must be added the names of Reinhold Baer (at Illinois and Yale), Waldemar Trjitzinsky (at Illinois), and Antoni Zygmund (at Chicago). Many other mathematicians—even if they did not have large numbers of women students during the forties and fifties—consistently supported women's mathematical ambitions and willingly supervised women's Ph.D. theses. Among these were Paul T. Bateman (Illinois), Robert Cameron (Minnesota), E.W. Chittenden (Iowa), E.P. Lane (Chicago), C.C. MacDuffee (Wisconsin), R.L. Moore (Texas), and R.L. Wilder (Michigan).

Mathematical Specialization and Gender

Do women have particular aptitude for, or are they particularly attracted to, certain subfields of mathematical inquiry? The answer is by no means clear. In their study of the pre-1940 American women mathematics Ph.D.'s, Judy Green and Jeanne LaDuke have determined that the most popular fields of dissertation research were geometry (38 percent), analysis (30 percent), and algebra and number theory (24 percent). They attribute the preponderance of women in geometry to a handful of highly influential thesis advisers in that field who were especially receptive to women students. In general, Green and LaDuke argue against the notion of a "woman's field" within mathematics (1987: 20–21).

In recent decades, however, studies and anecdotes about women, men, and mathematics have put forward the view that women show a significant preference for discrete mathematics over analytical fields, while for men the preference is generally reversed (Hutchinson 1977; Helson 1971; Luchins and Luchins 1980). If there is any truth to the assertion that women prefer algebraic over analytical fields, this preference was not evident in the forties and fifties. Such a preference—if it exists at all—must be a relatively recent (post-1960) phenomenon.[9]

During the period 1940 to 1959, over 40 percent of the women earned their Ph.D.'s in the fields of analysis and applied mathematics, outnumbering those working in algebra and other discrete fields by a margin of over two to one. In particular, women seemed to be particularly attracted to relatively new fields in mathematics—applied mathematics, statistics, logic, combinatorics. This observation is consistent with the model put forward by Margaret Rossiter (1978), who asserts that in "a rapidly growing field, with a shortage of highly qualified people, women [are] tolerated and even sought out." Sadly, however, Rossiter's model also suggests that women are more likely to persist in fields that are out of fashion, or in which future progress is unlikely, because they are "more willing than men to endure the hardships of a stagnant or shrinking field" (147).[10]

An Introduction to the Women Interviewees

Nearly all of the thirty-six women interviewed for this study have maintained a lifelong connection to the mathematical and academic communities and have had careers in teaching, research, government, educational administration, or some combination thereof. Appendix B lists each interviewee alphabetically, together with information on when and where the Ph.D. was earned, dissertation adviser, general field of dissertation research, marital status, career pattern, and location and length of each interview. Regrettably, the group of interviewees does not include any of the more than twenty Roman Catholic nuns who earned Ph.D.'s from 1940 to 1959; many of them are no longer living, while a number of others declined to be interviewed.

While several of the women with whom I spoke have achieved a modicum of fame and even celebrity in the mathematical community, a good many have lived productive lives in comparative obscurity. What all of these women have in common is that each had the talent, the opportunity, the perseverance, and the good fortune to complete a Ph.D. in mathematics and that each has faced the challenge of making mathematics an important, ongoing part of her life.

Just over a third of the interviewees were either immigrants or children of immigrants from eastern and central Europe. Each of these women found her personal outlook irrevocably shaped by the emigrant experience. Many of the women in this group express a sense of gratitude toward America, accompanied by a strong work ethic and an intense ambition to succeed in their adopted country. Moreover, as the psychologist Ravenna Helson has suggested in her research on women mathematicians in the late 1950s, their European backgrounds enabled them to "avoid some anti-intellectual influences of the mainstream of American culture" (1971: 218).

Well over half of the women interviewed grew up in or near large North American cities; in particular, eight of the interviewees spent the majority of their childhood years in and around New York City. Their urban upbringing exposed them to a diverse cultural, intellectual, and political life. By contrast, six of the interviewees were raised in the urban or rural South, in families that prized education, especially education for women.

As explored further in chapter 3, the future career success and satisfaction of these women seemed to depend crucially on having had a family environment in which education was valued and in which their ambitions were supported and encouraged. Several of the women described home environments where intellectual and social issues were freely discussed and intellectual curiosity was actively nurtured.

The educational backgrounds of parents varied widely. A handful of fathers held Ph.D.'s or worked in the medical or legal profession; four of the mothers held doctorates. In particular, two fathers and one mother held the Ph.D. in mathematics. Several of the women reported having particularly well-educated mothers. In a few families, higher education was a long-standing tradition for both men and women: Mary Dean Clement reports that both of her paternal grandparents had college educations; Mary Ellen Rudin and Joan Rosenblatt report that this was true of both of their grand*mothers*.

In many cases, however, the parents had little formal education and were largely self-taught. As a consequence, they had tremendous respect for education and, because they had not had much themselves, prized it for their children, both male and female. The support and encouragement of at least one parent seems to have been of signal importance to a young woman's later development as a mathematician. In particular, a solid grounding of support from at least one adult—a parent or a teacher— at an early age helped to provide a bulwark against discouragement or discrimination farther down the road.

Many of these women identified an interest in mathematics as young girls, but this interest was often no more intense than other intellectual interests at the time. Contrary to the dictates of the myth, it was not so important that mathematical interest and talent emerged early; rather, it was much more important that *all* talents and interests be allowed to rise to the surface at an early age and that none in particular be suppressed. A few of the women showed an early interest in the sciences—particularly physics—but many others indicated that languages fascinated them even more. Not surprisingly, perhaps, many of the interviewees aspired to become teachers from an early age.

Half of the women interviewed earned their undergraduate degrees at coeducational institutions—liberal arts colleges, large state universities, and privately or publicly funded urban universities—in the United States. Five of the women earned their first university degrees abroad. Just over a third of the women interviewed completed their undergraduate study at single-sex institutions: eight at private women's colleges, three at coordinate colleges, and two at teachers colleges (which were effectively, if not explicitly, single-sex institutions at the undergraduate level). A substantial majority of the women lived at home, for financial reasons, for all or part of their college years.

Most of the women identified mathematics as their major early in their undergraduate studies. For those whose major was undecided at an early stage, nearly all had included mathematics among the possibilities, with the exception of Tilla Weinstein, who began as a major in English and philosophy and began to consider mathematics as an option halfway through her college education.

Nearly all of them identified an inspiring teacher, or teachers, at the high school or college level who particularly sparked their interest in mathematics. In many cases, college teachers encouraged these women to continue to graduate school, often recommending the same graduate school they had attended. For example, all three of the women interviewed who earned doctorates at the University of Chicago in the forties were encouraged to do so by Chicago-trained faculty at their undergraduate colleges. But in several other cases, the decision to attend a particular graduate school was made out of financial necessity or geographic convenience.

Many of the women experienced significant delays between receipt of the undergraduate degree and receipt of the doctorate; the mean elapsed time between receipt of the undergraduate degree and awarding of the doctorate is nearly eight years. Many women who attended college during the Depression simply did not have the financial means to continue directly to graduate school. Moreover, some women who took high school teaching jobs right out of college decided after a few years that they would rather teach more advanced mathematics to older and more mature students and thus went on to graduate study in mathematics after a hiatus of several years.

For the European emigrant women, delays between undergraduate and graduate study were often caused by the extreme adversity of moving from country to country. Quite a few of the interviewees, particularly those who earned Ph.D.'s in the 1950s, married before or during graduate school. For these women, family responsibilities, or the precedence granted to a husband's education or career, sometimes caused delays in their own educational progress. Moreover, a few of the women held full- or part-time jobs in graduate school, which sometimes delayed their progress to the doctorate.

The process of selecting an adviser and a field of dissertation research was not much different than it is now. Typically, a beginning graduate student would enroll in a course and become interested in the subject matter and the teacher at the same time; working with a particular adviser in a particular field would then be a natural outgrowth of having taken the course. Then as now, it was sometimes difficult to change advisers once a relationship had been established. A reluctance to change advisers could have serious consequences down the road for women who found themselves trapped in a comparatively unfashionable or unproductive line of research; in particular, this seems to have been a problem for the students of E.P. Lane at Chicago.

At many different universities, a high degree of camaraderie existed among the graduate students and, often, between the graduate students and faculty. At the same time, however, some obvious social distinctions were made between male and female graduate students. For example, at the University of Chicago during the war, women graduate students were responsible for preparing, serving, and cleaning up after the weekly colloquium tea, while men were generally exempted from any such responsibilities—a division of labor under which many of the women students chafed.

During the growth and expansion of the mid- to late forties, there was a mood of excitement on campus, and both women and men "caught the bug" of mathematical research during these years. For example, Jane Cronin Scanlon, Margaret Marchand, and Barbara Beechler paint especially vibrant pictures of the intellectual excitement at their respective campuses—Michigan, Minnesota, and Iowa—during the years immediately following the war. By contrast, Augusta Schurrer, who worked on her Ph.D. at Wisconsin from 1945 to 1952, noticed that the graduate school environment gradually became much more competitive and far less welcoming to women as increasing numbers of men arrived for graduate study.

Marital Status and Children

In their interesting and provocative study, *Women of Academe,* Nadya Aisenberg and Mona Harrington (1988) assert that American women have been presented with two mutually conflicting models of how to achieve personal and professional fulfillment, which they refer to as *the marriage plot* and *the quest plot.* Particularly in the 1950s, the marriage plot was the cultural norm for American women, who were expected to stay at home raising children and maintaining a supportive home for their husbands; the quest plot was the norm for men, who journeyed out into the wider world and engaged in public life.

The women mathematics Ph.D.'s of the forties and fifties were strongly affected by the conflict between the marriage plot and the quest plot. Fully three-quarters of the interviewees—twenty-seven of the thirty-six— have been married at one time or another during their lives. Moreover, most of these women embarked on their married and professional lives nearly simultaneously: only four of those who married did so more than five years after completing the Ph.D.

Among those women who received Ph.D.'s in mathematics prior to 1940, sixty-three percent never married (Green and LaDuke 1987: 21). Thus, the generation of women who earned Ph.D.'s in mathematics in the forties and fifties is really the first in which substantial numbers set out to challenge the cultural norms and interweave elements of the marriage and quest plots into the fabric of their lives. It is no simple matter to gauge their success. Of the twenty-seven interviewees who married, five were divorced at the time of the interview; three had divorced and remarried; and one had been divorced, remarried, and subsequently widowed.

Eighteen of the interviewees—two-thirds of those who married—have had one child or more. In keeping with the mores of the times, responsibility for children fell disproportionately on the women. Almost all of the women with children reported that they engaged the services of nannies, housekeepers, or babysitters at least some of the time and that such hired help was frequently hard to find. Mary Ellen Rudin reports that, at times, the expenses of child care amounted to more than what she was paid as a part-time teacher at the University of Wisconsin.

Fully half of the women interviewed were married for a significant period of time to another mathematician or to an academic in another field. In many of these couplings, a shared love of mathematics provided a strong social, intellectual, and emotional bond. But in nearly every academic marriage, the man's career development was given priority, and in some cases the woman was handicapped in her ability to finish the Ph.D. or secure a professional position commensurate with her abilities. Antinepotism rules remained in force at many American colleges and universities through the late 1960s and greatly limited the career opportunities of many of the women in this group.[11] In addition, some of the interviewees reported difficulties associated with continuing work or study while pregnant, particularly in the 1950s; it was considered inappropriate and even obscene for an obviously pregnant woman to hold forth in front of a classroom, especially before a class of men.

Among the interviewees in the present study, nine remained unmarried throughout their lives; yet another decided to stay single after a brief, early marriage ended in divorce. Among the unmarried, few expressed regret or bitterness; for some it was clearly a choice. It is interesting to note that seven of the women who never married received their Ph.D.'s prior to 1949; their undergraduate education had mostly taken place in the 1930s. They came of age at a time when professional women were largely *expected* to remain unmarried; marriage, therefore, may have seemed less compulsory to them. Their female teachers—and primary role models—had for the most part been single women.[12]

Married or not, all of the women interviewed were affected by the cultural norm of the marriage plot. Almost without exception, they reported that as children they *expected* to marry and have families when they grew up. A few of the women who formed professional ambitions at a young age defiantly resolved that they would *never* marry, but these resolutions were sometimes broken later on. Both Janie Bell and Dorothy Maharam Stone made such resolutions during their teenage years, yet each married a fellow mathematician shortly after completing the Ph.D.

Academic Career Paths

Of the thirty-six women interviewed, twenty-eight were eventually tenured professors at one or more colleges or universities in the United States. Of these twenty-eight, six achieved tenure at women's colleges (Schafer, Anderson, Asprey, Bates, Freitag, and Beechler). One (Marchand) was tenured at a small coeducational liberal arts college, Adrian College in Michigan. Five ultimately received tenure at comprehensive state universities (Willerding, Granville, Steinberg, Schurrer, and Williams). Remarkably, the remaining sixteen were ultimately tenured at research universities.[13]

While the majority of the interviewees ultimately held tenured positions at colleges and universities, it was unusual for a woman to start out on the tenure track and remain on it continuously throughout her career. In fact, only nine of the twenty-eight women in the ultimately tenured group held tenure-track positions continuously from the Ph.D. onward. Of these nine, seven worked for one institution for essentially their entire career (Anderson, Asprey, Bates, Wurster, Larney, Schurrer, and Freitag). Several of the other nineteen women held temporary faculty positions, or experienced periods of unemployment, during the first several years after receiving the doctorate, most commonly because they were starting a family or deferring to the career of a spouse. A few of these women spent extended periods during the early years of their careers in nonacademic employment (Eberlein and Pless).

What was the academic job search like for new Ph.D.'s in mathematics in the forties and fifties? Nowadays, college and university mathematics departments advertise their open faculty positions in the journals of the professional organizations for mathematics—such as AMS, MAA, and the Society for Industrial and Applied Mathematics (SIAM, founded 1952)—and job candidates make direct application for these positions. But in the 1940s, new Ph.D.'s generally landed their first jobs almost exclusively through an informal system that I refer to as *the old-boy/old-girl network*. Typically, a student's adviser would put the word out to colleagues and friends at other institutions that he had a student who was completing a Ph.D. and about ready to take a teaching job. At the same time, department heads seeking new faculty members would contact graduate departments where they had contacts or connections in search of fresh Ph.D.'s. Through these informal conversations, new Ph.D.'s were matched with job openings with remarkable efficiency.

One of the predictable results of the old-boy/old-girl system was that women Ph.D.'s ended up in the mathematics departments of women's colleges, many of which were headed by women Ph.D.'s of the pre-1940 generation. At the women's colleges, teaching loads were often very high and research opportunities were limited. At the same time, the mathematics departments at coeducational colleges and universities, headed and for the most part staffed by men, generally recruited men. If a woman wanted to continue in research, she had to take active control of her job search, often moving through a succession of positions until she found one in which her research efforts were supported and rewarded.[14]

Four of the women interviewed *never* held a regular tenure-track job until they were abruptly appointed to tenured positions when they were already relatively far along in their research careers. From 1955 until 1967, Jean Rubin held temporary faculty jobs, first at the University of Oregon and later at Michigan State. In 1967, however, both she and her husband, also a mathematician, were offered tenured faculty slots at Purdue. Only then did Michigan State come through with a counteroffer of tenure, but the Rubins chose to move to Indiana nevertheless. Lida

Barrett, who married a fellow mathematician in graduate school, was repeatedly denied graduate support and regular faculty positions because of antinepotism rules while her husband was alive. But on his death in 1969, she was immediately offered a tenured professorship at the University of Tennessee. After ten years of temporary faculty status at the University of Wisconsin, Mary Ellen Rudin was appointed a full professor with tenure in 1970. Vera Pless received her first regular faculty appointment in 1975, when the University of Illinois at Chicago hired her as a tenured professor of mathematics and computer science.

Sudden promotions and tenurings of women became increasingly common in the early 1970s, particularly in the aftermath of Title IX, as colleges and universities were scrambling to find highly qualified women to promote or appoint to tenured positions.[15] Title IX of the Educational Amendments Act of 1972 is the piece of federal legislation that effectively "banned sex discrimination in any program of an institution [of higher education] receiving federal funding, including sports, textbooks, and the curriculum" (Rossiter 1995: 382). It was the culmination of more than ten years' organized effort on behalf of women in academia and government seeking equal pay, equal treatment, and equal status.

Discrimination and Opportunity

Without exception, the interviewees all experienced significant obstacles to the pursuit of careers in mathematics because they were women. Particularly in the 1930s and 1940s, many institutional barriers were in place for women who wished to study mathematics. In both high school and college, for example, Grace Bates had to request permission to take advanced mathematics courses, normally available only to men. At all educational levels, women regularly encountered faculty and administrators who either blatantly or subtly conveyed the message that women couldn't or didn't do mathematics. When she entered MIT as an undergraduate at the age of sixteen in 1937, for example, Domina Spencer was denied a full-tuition scholarship on the ground that women—who would eventually marry and drop out of sight—were "a bad investment." She, like many of the other interviewees, refused to back down in the face of such overt discrimination and simply resolved to prove them wrong.

As has already been mentioned, these women were at various times adversely affected by antinepotism rules and discriminatory attitudes toward pregnancy. Unmarried women did not always fare better: when Margaret Willerding came to Washington University as an instructor in 1947, she was told that she would not be promoted as quickly as her male colleagues. Many other women knew that they had been passed over for promotions and raises in favor of men, but few were informed of the situation so bluntly.

Some of these women faced multiple forms of discrimination. Evelyn Granville, who as a young girl aspired to a career in teaching, spent nearly fifteen years working in government and industry, where the opportunities for an African American woman were greater than in academe. She did not return to university-level teaching until 1967. In the case of Rebekka Struik, gender discrimination merely compounded other forms of discrimination she experienced. It was her left-wing political views and activities that left her vulnerable to attack during the McCarthy years of the 1950s. By virtue of her involvement with leftist student groups, she lost a job as a computer programmer at the University of Illinois and also lost a fellowship to attend graduate school in mathematics at Northwestern University.

All of the women interviewed had to struggle against discrimination to a greater or lesser extent, but they differ widely in their attitudes toward such discrimination. Some of them feel that their gender is an asset that has, at times, provided them with certain advantages. Even before the days of Title IX and affirmative action, some individuals took a special interest in the promotion of women. Kenneth Wolfson, the forward-looking chair of the mathematics department at Rutgers University, built up a first-class research faculty in the sixties and seventies by mining untapped sources of talent, hiring Jane Cronin Scanlon and several women Ph.D.'s of her generation in the process.[16] Federal programs, particularly at the NSF, provided opportunities and resources for women to pursue graduate education and postgraduate research. Tilla Weinstein, for example, attended graduate school at NYU from 1955 to 1959 on an NSF graduate fellowship that provided living expenses for herself, her husband, and her child.

In the late 1950s and early 1960s, organizations as diverse as the National Research Council (NRC), the American Association of University

Women (AAUW), and the American Association of University Professors (AAUP) began a concerted assault on antinepotism rules at U.S. colleges and universities (Rossiter 1995). Over the course of a decade, these rules gradually disappeared. The women's movement of the mid- to late sixties—given impetus by the 1963 publication of *The Feminine Mystique* by Betty Friedan—applied the necessary social and political pressure that eventually led to government action against sex discrimination in higher education.

With the federal legislation of the 1970s, colleges, universities, and the federal government were all required to provide equal opportunity for women, and special efforts were made to recruit women to posts that had hitherto been inaccessible to them. While many of the resulting affirmative action efforts were carried out grudgingly, many more opportunities opened up for women in the 1970s and 1980s. Many of the women mathematicians in this study—just entering the peak of their professional creativity and energy—were able to take advantage of these new possibilities. They were invited to serve on national boards and hold national office in professional societies; they were promoted to tenured professorships and rewarded with long-overdue gender equity pay increases; they were invited to move into influential administrative positions.

Many of the women interviewed were sympathetic with the goals of the women's movement, although some were alienated by its methods. Some of them fully embraced the arrival of the movement and became active in efforts to enhance the status and opportunities of women in mathematics and the sciences—most notably Alice Schafer, who played a key role in the founding of the Association for Women in Mathematics (AWM). Established in 1971, the AWM has done more than any other organization to increase the visibility of women in mathematics and to advocate for women's equality in the profession. Several of the other women interviewed are or have been members; some are actively involved in the organization, others not.

Deflection—and Defection—from Academic Careers

Of the eight women who were not eventually tenured in mathematics or mathematics-related departments, four can be described as having been *deflected*—to borrow a term from Aisenberg and Harrington (1988)—

from their original plan to pursue a career in college or university teaching and/or research in mathematics. For three of the women (Bell, Calloway, and Morley), marriage to a graduate school classmate sharply curtailed their pursuit of an independent academic career. Mary Dean Clement, who never married, held a succession of unsatisfying teaching positions in the forties and early fifties and ultimately went to work at a military research facility attached to the University of Chicago, where she had earned her degree.

In addition to these four women who were deflected from the academic career track, five other interviewees can be said to have *defected* from academia. Margaret P. Martin actually held academic positions in statistics for over twenty years and was tenured in two of them. As her parents grew older, however, she chose to live closer to their home in St. Paul and left academia to work as a biostatistician for the Forest Service, where she spent the last twenty or so years of her career. Her defection came not from disappointment or disaffection but rather for personal reasons.

By contrast, Joan Rosenblatt, Betty Jane Gassner, and Susan Hahn never *sought* academic positions on completion of their doctorates. Rosenblatt, whose parents were professors at Columbia, was already quite familiar with academic life and chose instead to pursue a career with the federal government (NBS/NIST). Gassner and Hahn, who had been graduate students at NYU, chose to remain in the New York area; Gassner worked in a succession of industrial research positions, while Hahn spent over twenty-five years at IBM.

Jean Walton is unique among the women who did not end up as professors of mathematics. During the hiatus between her master's degree and work toward the doctorate, Walton served as assistant dean at her undergraduate alma mater, Swarthmore College. On completing the Ph.D. in mathematics at the University of Pennsylvania in 1948, her stated objective was to teach mathematics at a small liberal arts college. But on the strength of her prior administrative experience, she was offered—and accepted—the position of dean of women at Pomona College, where for several years she also taught classes in the mathematics department. As the years went by, she became more deeply committed to her administrative work and stopped teaching mathematics altogether, although she maintained her affiliation with the mathematical societies. Ultimately, however, she severed her connection to the mathematical community entirely.

Varieties of Mathematical Activity

With the exception of Jean Walton, all of the women interviewed identify themselves as mathematicians or as practitioners (*doers*) or teachers of mathematics. Most regard teaching and research as the activities that are most crucial to the ongoing growth and development of the field. Bearing in mind that any attempt to classify these women in terms of their primary career activity is but a rough approximation, I offer three general categories that describe the main focus of their mathematical activity and their identity as mathematicians: researcher, teacher, and scholar-teacher.

Nine of the women interviewed are readily identifiable as researchers: Maharam Stone, Spencer, Cronin Scanlon, Rudin, Morawetz, Eberlein, Rubin, Pless, and Weinstein. Yet another, Joan Rosenblatt, pursued a varied career in the federal government that included lengthy administrative stints; yet she, too, seems to identify primarily with research and scholarly activity. In addition to these ten, Lida Barrett's primary focus was in research until she became involved in academic administration in the early 1970s.

The high-status role of researcher, which was presented as normative to most doctoral students in the postwar years, was the role most at odds with the cultural expectations of women, particularly during the 1950s. It is interesting to note, therefore, that all eleven of the women who identified as researchers are or have been married for a significant part of their active professional lives and that only one of these women was childless. This raises the possibility that married women with children may in fact have been more acceptable in the male-dominated mathematical research community because they had also fulfilled their social obligations as women. It also raises interesting questions as to just how effectively unmarried or childless women were integrated into the postwar American mathematical community.

Nine of the women were associated with the more traditionally female role of teacher: Anderson, Wurster, Larney, Schurrer, Beechler, Schafer, Marchand, Granville, and Williams. Again, this classification is a fairly rough approximation; Marchand, for example, spent four of her first six postdoctoral years as a research biostatistician but from 1956 until her retirement in 1990 worked primarily as a teacher of mathematics.

Although Larney, Granville, and Wurster were involved in the writing of mathematics texts, teaching was clearly their primary focus.

For three of these women—Beechler, Schafer, and Marchand—the choice of teaching as a career clearly involved a turning *away* from research. From a very early age, Barbara Beechler was enthralled with the romance of scientific discovery, but her graduate school experiences were a source of alienation and disillusionment and soured her on the possibility of future research. Alice Schafer feels very strongly that the graduate faculty at Chicago in the forties simply did not take the research potential of women students seriously; in a sense, she feels she was limited to a teaching career from the start. Her anger over these experiences has been creatively channeled over the years into her enthusiastic mentorship of young women in mathematics and into her energetic leadership of the AWM. For Margaret Marchand, by contrast, the turn from research to teaching was a welcome return to her original ambition and so did not, apparently, engender the kind of bittersweet or angry feelings expressed by Beechler and Schafer.

Many of the women interviewed occupy a position intermediate between the categories of *researcher* and *teacher,* which I refer to as the *scholar-teacher.* For the scholar-teacher, scholarship and teaching are regarded as mutually interdependent, although teaching often takes preeminence. Asprey, Bates, Willerding, Freitag, Lax, Luchins, and Struik have all been scholar-teachers; each has pursued scholarly interests that at times proved indispensable to their teaching. For example, Winifred Asprey established close ties between Vassar College and IBM, which culminated in the establishment of a computer center at Vassar in the mid-1960s. Grace Bates, whose doctoral research was in abstract algebra, later conducted research in probability and statistics with Neyman at Berkeley. This dramatic shift of research field came as an outgrowth of her need to learn the subject in order to teach it at the undergraduate level at Mount Holyoke College. For both of these women, research and scholarship were intertwined with their teaching in a fairly seamless way.

In addition to the seven women already mentioned, the career of Margaret Martin prior to her move into the Forest Service had the quality of the scholar-teacher: the statistical research she conducted over the years arose from her role as a statistical consultant for schools of medicine and public health, particularly in connection with her work as a teacher of statistics to medical students. Maria Steinberg's professional experience defies neat categorization but has something of the spirit of the scholar-teacher. Research and teaching tended to be sequential rather than simultaneous activities during her career. Although her involvement with research effectively ended in the 1950s, she has maintained an ongoing sense of connection to the research community through her husband, Robert Steinberg, so that the spirit of mathematical research is very much a part of her sense of herself as a mathematician.

As is noted in chapter 1, many mathematics faculty at prestigious colleges and universities in the early decades of this century viewed themselves as scholar-teachers. Thus for the women mathematicians of the postwar generation, the role of scholar-teacher may be viewed as a return to the turn-of-the-century academic ideal. It may also be viewed as a creative synthesis of the "female" role of teacher and the "male" role of research mathematician, but for many of these women it was also a not entirely satisfactory compromise between two unacceptable career alternatives.

The other six women interviewed (Clement, Bell, Calloway, Morley, Gassner, Hahn) defy neat categorizations. Clement and Morley may be regarded as deflected scholar-teachers, Calloway and Gassner as deflected researchers. Susan Hahn's career at IBM was shaped by the commercial needs of the company and took her progressively further away from her mathematical origins. Although Janie Bell cannot be said to have had a career in teaching, she has held short-term and temporary teaching positions at every level from elementary school through college and very clearly sees herself as a teacher of mathematics.

Accomplishments and Expectations

Any attempt to assess the degree of success that these women have achieved in their lives and careers must take into account the extent to which their accomplishments fulfill the expectations they had for themselves. For example, many of the women interviewed whom I have categorized as teachers have nevertheless expressed tension or dissatisfaction over the road not taken in research. Some of them genuinely wish they had been granted the opportunity to prove themselves as researchers; others simply report feeling a tension between the research expectations others had for them and the professional activities they freely chose to pursue. Violet Larney goes so far as to raise the provocative question: are women actually *capable* of conducting research in mathematics at the same level as men?[17]

Success and satisfaction are derived from many things: a feeling that one has contributed substantially to the body of mathematical knowledge; a feeling that one has made a contribution to the welfare of students or to the mathematical and academic communities as a whole; the sense that one has made the best use possible of one's talents and energies. The chapters that follow explore in detail the lives and careers of this special group of thirty-six women in an attempt to better understand the personal and professional dimensions of success and failure, satisfaction and frustration, in their lives and careers, in and around mathematics.

3

Family Background and Early Influences

Nature, Nurture, and Family Niche

Much of what we currently know about the early lives of women mathematicians comes from a psychological study conducted in the late 1950s by Ravenna Helson and her colleagues at the Institute for Personality Assessment and Research at the University of California at Berkeley (Helson 1967, 1971). At that time, it was estimated that there were approximately three hundred women mathematics Ph.D.'s in the United States—including nearly all of the women who earned mathematics Ph.D.'s in this country in the 1940s and 1950s. Helson's work is interesting and important for many reasons, not the least of which is the considerable overlap between her subject pool and the women I consider in the present study. Indeed, many of my interviewees identified themselves as participants in Helson's project.

Helson's goal was to understand "the creative personality" in mathematics and in particular to ascertain the distinguishing characteristics of creative women mathematicians. She made a careful study of forty-five women mathematicians, eighteen of whom were considered to be creative, a designation based solely on the evaluation of (male) mathematicians at distinguished research universities. The remaining twenty-seven women mathematicians served as a comparison group. This method of identifying creative women in mathematics is in itself problematic, as it clearly reflects the judgment of the male majority, as well as the preeminence of mathematical research over other forms of scholarship and professional activity.

Helson's study nevertheless provides valuable insight into the psychological development of those women who met the highest standards of the American mathematical community in the 1950s. The profile that emerges is that of a girl growing up in a family with relatively few brothers and a dominant but emotionally distant father, whose intellectual and educational attainments differ sharply from those of his wife. As the girl grows older, she comes to identify strongly with the father, particularly with regard to intellectual pursuits; somewhat introverted and socially isolated, her approach to life is flexible and nonconformist. The characteristic that describes her best is "rebellious independence" (1971: 214–218).

While this profile may describe a significant minority of the women mathematics Ph.D.'s of the forties and fifties, it offers just one model of how girls of this generation grew to become active participants in the American mathematical community. In fact, they came from a variety of backgrounds and family circumstances, and each followed her own unique path to mathematics. What they all have in common is one central act of "rebellious independence": the crucial decision to set out on that path.

In his recent, provocative study, *Born to Rebel,* Frank Sulloway (1996) offers the concept of the *family niche* as a means of explaining why and how individuals emerge from their families of origin to lead lives of originality and nonconformity. As children, each of the women interviewees carved out a unique family niche within which her interest in mathematics began to grow and develop. Some of these girls came from warm families where their emotional and material needs were met and their intellectual interests were actively encouraged. Others came from more adverse circumstances in which their interests developed in spite of disparagement, or as a defense against emotional or material difficulty, or as a crucial means of self-expression. What the stories bear out, in their similarities and in their differences, is that mathematical ambition arises from a complex interplay of factors: some genetic, some familial, still others external to the family. Because the family of origin is so crucial to personality formation, it is there that I begin.

Mothers and Fathers

Thirty-one of the thirty-six women interviewed grew up in a traditional family structure in which the mother and father were married and present in the home. Of the remaining five, three (Maharam Stone, Schafer, Bates) experienced the early death of one or both parents. Domina Spencer's father was a traveling salesman, and as such saw his wife and children only at irregular intervals during Domina's childhood. Domina, her sister Vivian (twelve years older), and their mother moved from place to place as Vivian, who also became a professional mathematician, attended college at Oberlin and graduate school at the University of Pittsburgh and the University of Pennsylvania.[1] Mr. Spencer would join the family in these various locations when work and finances permitted. Evelyn Granville was the only woman interviewed to experience the marital separation of her parents; she and her older sister were raised by her mother, but her father remained a vital presence in their lives as they grew up.

Nearly all of the women interviewed came from family backgrounds in which education was valued very highly. As has already been noted, many of the women were immigrants, or children of immigrants, from central and eastern Europe. These immigrant families brought with them to America the respect for learning and culture that was the hallmark of many European societies and, in particular, a central feature of the Jewish tradition. In many of these families, the myth of a better life in a land of opportunity became an extremely powerful motivating force for both parents and children, spurring them to achievements in work and at school.

Twenty of the thirty-six interviewees had at least one parent who had attended some college. In two cases, both parents had earned doctoral degrees: both of Rebekka Struik's parents had Ph.D.'s in mathematics; Joan Raup Rosenblatt's mother had a Ph.D. in economics, while her father (a student of John Dewey) had his Ph.D. in educational philosophy and psychology.[2] Four of the women had one parent who held a doctorate (or equivalent degree awarded in Europe): Maria Steinberg's mother held a Ph.D. in history; Vera Pless's mother earned a dentistry degree in Russia and, on her arrival in America, went back to school for a D.D.S.; Cathleen

Morawetz's father was the Ph.D. mathematician J.L. Synge; Anneli Lax's father had a medical degree and worked as a surgeon and a urologist.

For the women whose *mothers* had doctorates, their mother's educational and professional experiences made a profound and lasting impression. Joan Raup Rosenblatt's mother worked as a professor of economics at Barnard College; while she was pregnant with Joan, she was "the first woman on the faculty of Barnard College to get a maternity leave—unpaid, but with the promise of getting her job back" (Rosenblatt: 2). Vera Pless, an only child, could not help but be aware of her mother's dental practice: "She had an office in the house when I was a child, so that she could watch after me" (Pless: 2). Rosenblatt and Pless grew up in environments where the combination of motherhood and professional work, while decidedly unusual, nevertheless seemed natural and even desirable.

By contrast, Rebekka Struik's mother never really had satisfactory employment as a mathematician, and Rebekka was always acutely aware of her mother's conflicted feelings: "While I was growing up, she didn't do any mathematics, and I saw her as very unhappy . . . being primarily a housewife" (Struik: 3). Similarly, Maria Steinberg's mother was never employed as an historian but seemed to have far fewer regrets. Her intellectual attainments made an enduring positive impression on Maria, who recalls with some fondness, "She always was interested in things of the mind" (Steinberg: 2).

For exactly one-third of the women interviewed, a fairly dramatic educational disparity existed among the parents. In seven of these families, the mother was more highly educated than the father (Spencer, Martin, Asprey, Steinberg, Larney, Barrett, and Pless); in the remaining five families, the father was more highly educated than the mother (Maharam Stone, Walton, Morawetz, Lax, and Weinstein). Curiously, in the majority of these families, the *mother* was the parent who was most adamant that her daughter be educated. If the mother was the parent with the superior educational attainment, she usually wanted her daughter(s) to have the same educational opportunities that she'd had. But if the mother was less highly educated than the father, she had nevertheless had the opportunity to observe firsthand the advantages of education. In some cases, she deeply regretted not having had more education of her own and thus sought for her daughter what she had been unable to secure for herself.

Winifred Asprey and Lida Barrett provide illustrative examples. Asprey's mother was a graduate of Vassar College, devoted to her alma mater; Barrett's mother was a 1912 graduate of the University of Texas and a firm believer in college education. Both fathers dropped out of school—Asprey's in the tenth grade, Barrett's in the eighth—for financial reasons to help support their families. Yet both fathers went on to become reasonably successful businessmen; each is aptly characterized as a self-made man.

Winifred Asprey's father was an ambitious man with intense intellectual and political interests. After dropping out of school, he worked first as a jeweler and then as a candy jobber, staying in business during the Depression by accepting barter as well as cash. Ultimately, he left business altogether and entered city politics in their home town of Sioux City, Iowa. Asprey recalls:

I never saw Dad read anything that was trash. It always was history and biography: he loved it. . . . He knew more about current events, he knew more about history and business matters, than I think I would ever know. And I don't think that we ever differentiated at all in our minds that, you know, Mother was "college" and Dad was not—except that we knew that it was true. He was a very intelligent man, and a very kind man. He supported loads of families in the city, and some that he'd known since childhood. And everybody knew him. Everybody liked him. (Asprey: 2, 4–5)

Lida Barrett's father was ingenious and inventive, with a great deal of practical intelligence. The signal event that shaped his adult life occurred while he was still a teenager: he lost a leg in a work-related railway accident. He was fitted with an artificial leg, and some years later went on to purchase the company that supplied his prostheses, the Texas Artificial Limb Company. Barrett remembers him as someone who was "very good at numbers, really enjoyed numbers and games and things, and was very intelligent," although she adds, "I don't have a sense of how well-read he was" (Barrett: 3).

In the twenties, thirties, and forties, it was still possible—if by no means easy—for a man to make a living without having had a lot of formal education. But in those days it was far more difficult for a woman to do so. Thus it comes as no surprise that the *mothers* of Winifred Asprey and Lida Barrett proved to be the most consistent and vocal advocates for the college education of their daughters. Asprey—the eldest of three children, with two younger brothers—recalls that her mother took an extraordinarily active interest in her education, regaling her with stories of Vassar College from a very early age. Barrett, who grew up in Houston with an older brother and a younger sister, asserts unequivocally, "My mother was the one that was determined we were all going to have college educations" (Barrett: 3). Indeed, both Barrett and her sister attended Rice Institute (now Rice University), where tuition was free in the 1940s.

Neither of Domina Spencer's parents completed high school, but her mother had considerably more education than her father did. Her father "had a third grade education and thought that was plenty!" At the same time, he maintained a lively intellectual life. "He read continuously—history, politics—and wrote poetry continuously," Spencer recalls. Whenever he wrote to his family, his mailings included a one-page letter and a one-page poem. She likens his work to "Edgar Guest poetry" and believes that "he would have made a very good writer" (Spencer: 4).

For Mrs. Spencer, however, the situation was very different:

Mother wanted to be an artist, and she was very much interested in music also. But what she wanted most when she was a girl was to go to Europe to study art and go to college here. Her family wouldn't let her. She went to a business college. That was it. I don't believe she'd even completed high school because [of] an incident over an epidemic. Mother was going to public school, [and] when she was probably around twelve, something around that, the state confiscated all her books and burned them because there was an epidemic. And my grandfather was a stubborn German, and he thought it was wrong to take those books that he had bought and burn them. [The school] wouldn't replace them, so he wouldn't let Mother go back to school! She was at the top of her class all the time she went to public school. She was the brightest student in the class, and suddenly, she wasn't allowed to go to school at all. (Spencer: 4)

The unfairness of having been deprived of an education of her own—particularly in view of the fact that her older brothers *had* been allowed to finish school—led Domina Spencer's mother to a passionate insistence on the education of her daughters. At the same time, however, she had serious reservations about the quality of the public schools, and so she undertook the education of her daughters at home.[3]

The experience of Tilla Weinstein illustrates how a mother who has appreciably *less* education than her husband can play a crucial—if inadvertent—role in her daughter's education. Weinstein's father emigrated from Russia to New York City at the age of thirteen. He graduated from City College and Harvard Law School and established a law practice in New York City. Her mother—whom she describes as having "almost graduated from high school; her typing teacher refused to let her graduate"—worked as a legal secretary prior to marriage and assisted her father in his law practice afterward. "I came as close to killing my mother in the process of innocently being born as one can," Weinstein recalls, "and she was literally unable to have children after. But also, they couldn't have afforded it. They could barely afford having me" (Weinstein: 2).

Weinstein's parents' marriage was stormy, and her father was seldom at home. "For me," Weinstein says, "growing up was very much being with my mother, and my father wasn't much of an influence at all. . . . It was a happy house when he wasn't there." As an only child, Weinstein was her mother's pride and joy, the center of her universe: "She spoke to me at great length about absolutely everything. . . . I loved listening to her and she loved listening to me. And I imagine that just being taken as seriously as I was by my mother was very important" (Weinstein: 2).

Weinstein's mother did not encourage her daughter to pursue advanced education but, rather, encouraged her to be whatever she wanted to be. This encouragement gave Weinstein the sense of freedom, security, and self-confidence that enabled her to continue her education as far as she could go. Her mother seems to have been surprised at the outcome:

I think she never understood the implications of what she admired and encouraged in me. She would have wanted me to excel at everything and marry someone who could give me a very comfortable life financially, so I could stay home and raise her grandchildren. . . . She never quite understood how things turned out the way they did! She didn't see the connection between encouraging me in every way and my not stopping. (Weinstein: 3)

There were, of course, numerous ways in which *fathers* exerted their influence to ensure the education of their daughters. In families of daughters, younger daughters were often treated as if they were sons—either covertly, as in the case of Jean Walton, or overtly, as in the case of Barbara Beechler.

Jean Walton was the fourth of five daughters born to George Walton—headmaster of the Quaker-affiliated George School outside Philadelphia—and his wife. Again, there was an educational disparity between her parents. Her mother had been educated through two years of normal school and was a teacher before she married; then, as was customary at the time, she left teaching to devote her full-time energies to home and family. Her father had a master's degree from the University of Pennsylvania, and though he aspired to earn more advanced degrees, he never did so. She characterizes her childhood relationship to her parents in the following way:

> I was very much more emotionally tied to what my father was thinking and what I might do to please my father, than I was with my mother. [But] I did not have an especially warm relationship with my father. He was a very loving—in words, and in manner, in a way—but he was not a warmly affectionate, person. He was busy, he was remote, he didn't talk very much. [Yet] somehow or other I had this strong need to please him—which I didn't recognize particularly at that point as being anything special. That came later. . . . I've in recent years checked this with my other sisters, and none of them had it this strongly. (Walton: 3)

Walton points out that she was "the last planned child" in a family of daughters. "Mother did confess to us that, indeed, the fifth one was an accident," she recalls. "There's no doubt that Dad would have liked a son" (Walton: 4). Perhaps her father, desirous of a son, communicated his disappointment to her; she responded with a tremendous urge to please him and perhaps to fulfill his educational ambitions as well. As it turned out, Walton was the only one among her sisters to continue her education to the doctorate.

Barbara Beechler was the younger of two sisters. Her sister Virginia, six years her senior, "was a homecoming queen; she was a very beautiful girl. So there was all this nonsense about how 'Virginia has the beauty and Barbara has the brains.' And we were very, very different" (Beechler: 3). Barbara's interests, and indeed her entire approach to the world, contrasted sharply with her sister's:

> I was very mechanically inclined—the typical little tomboy—and my father thought of me as his son, and I stayed with him all the time. But my mother and sister were constantly unhappy with whatever it was I was doing. That was bad news, you know; this insistence on baseball and running around and no dolls and so forth, was really bad news. . . . I was not my sister, and that was mentioned frequently. (Beechler: 2–3)

Beechler's father was a career employee of the Illinois Bell Telephone Company. Frustrated by the limits of his own education and by the difficulty of supporting his family during the Depression, his worries were compounded by ill health; he had been gassed during World War I and spent most of World War II in a tuberculosis sanatorium. It is clear that he sought a better life for his daughters and in particular firmly believed that both of them should have a college education. "My father was thought to be a fool by all of his colleagues at the office," Beechler recalls, "because he was educating his daughters" (Beechler: 2).

While both Beechler daughters did, indeed, complete college, it was Barbara whom he saw as the child who could fulfill his own dreams and aspirations. Her father's idealistic goals for her were consonant with her own interests and desires but manifestly at odds with her mother's practical and pragmatic view of how her daughters' lives should unfold:

My father was *inordinately* proud of me and insisted that I was going to be a scientist. And so, you know, from a very early age, I really *did* think I was going to be a professor of physics, long before I could pronounce the word *physicist!* Daddy thought that was really very important. My mother, on the other hand, I now realize, thought that was a terrible idea. I mean, I'm sure she worried terribly about this whole thing. She thought I would never finish going to school. She thought [I should] go out and earn an honest living, for heavens' sakes! (Beechler: 2)

As Barbara Beechler grew older, she ultimately developed much closer relationships with both her mother and her sister. What is more, she came to realize that her mother's decision to go to work during World War II, while her father was ill, was what made it possible for her to attend college at all. According to Beechler, it was her mother—despite all her misgivings—who "essentially put me through school" (Beechler: 10).

While it was often true that one parent took a more active interest in a daughter's education, the support of both parents—and occasionally the whole family—was usually what made a daughter's education possible. The upbringing of Edith Luchins provides a illustrative example of this. Luchins emigrated with her family from Poland to New York City when she was just six-and-a-half years old. Neither of her parents had had very much schooling in Poland, and her father felt the lack of formal education especially keenly: as a Jew in Poland, a college education had been all but closed off to him. Like the fathers of

Winifred Asprey, Lida Barrett, and many of the other women interviewed, Luchins's father was largely self-educated and had a lively intellectual life of his own making. Edith—as the oldest child and the one
most inclined toward academic work—became the academic standard-
bearer for the next generation:

I had a brother, but he was younger; and as a child he was not well, so he had
trouble doing the regular schoolwork. Then he went early into the Army, so he
wasn't going to be the one to carry out the educational aspirations of the family. . . . The whole family supported me and wanted me to succeed. Everybody
agreed: "Edith's the scholar!" So that was taken for granted. (Luchins: 6, 37)

Thus Luchins's entire family—her parents, her two younger sisters, and
her younger brother—came to salute and to support her academic goals
as they would the achievements of an eldest son.

Many of the women with whom I spoke had the good fortune to be
raised in families where *both* parents communicated to *all* of their children an infectious enthusiasm for learning and culture. Herta Freitag describes just such an environment in her vivid recollections of family life
in Vienna in the 1920s. Her father was a managing editor of *Neue Freie
Presse,* which she describes as "Vienna's foremost newspaper"; her
mother was an artist; and her brother Walter—who has had a long career
as a conductor with New York's Metropolitan Opera—was a musical
prodigy whose "talent was discovered when he was five years old." She
recalls her family life as one of shared intellectual and artistic passions:

What we specifically liked was to sit together—specifically, say, weekends—together, but everybody doing his thing. Father would almost invariably read the
newspaper, Mother would do her embroidery, Walter would compose, and I
would do mathematics. We even coined a German name for that—it was very,
very important for us: *Eine wonnige Gruppierung. Wonnig* is a rarely used
word—a delightful *Gruppierung*—group togetherness. (Freitag-1: 2)

Similarly, Joan Rosenblatt, both of whose parents had Ph.D.'s, describes
the environment of her childhood as "supportive, and stimulating" (Rosenblatt: 2). Like Edith Luchins, she was the eldest of four children, with
two sisters and a brother. Remarkably, she adds, "all of my siblings
went to graduate school, [although] I'm the only one who got a Ph.D."
(Rosenblatt: 3).

While most of the women interviewed had at least one supportive parent who valued education and achievement, two of them report that they

generally lacked this kind of support in their families growing up. Dorothy Maharam Stone was the fifth of six children in her family (three girls and three boys). Her father was a rabbi and her mother died when Maharam Stone was about five years old. Her family valued "the education of boys in rabbinical studies," but "the education of girls didn't matter" (Maharam Stone: 2). While her father did nothing to prevent her from going on in her education, she had to look outside the family for support and encouragement.

Similarly, Margaret Willerding recalls that her parents—neither of whom had graduated from high school—were not particularly interested in her education. But from a very early age, she resolved that she "was the kind of person that was going to go [to college] come hell or high water. . . . I was very ambitious" (Willerding: 2–3). Her parents, for the most part, adopted a laissez-faire attitude toward her ambitions. But when Willerding began work toward the Ph.D., her parents spoke up at last:

And that's when my mother said, when I said I was going to get my Ph.D., she thought that was rather silly. . . . I think my mother wanted me to get married, you see. And I wasn't the kind of girl my mother wanted. She wanted a social butterfly and she got a bookworm. But at least my father said, "If you want to do it, go ahead and do it." So I did. (Willerding: 3)

Sisters, Brothers, and Others

While parents have an important part to play in developing the talents and interests of their children, the environment of the family is a complex web of relationships that may also involve sisters, brothers, and significant others. In families with two or more children, interactions between siblings play a decisive role in forming the personality of each individual child. In families with only one child, however, the larger community often plays a key part in shaping her interests, values, and ambitions.

Ten of the interviewees were only children. Vera Pless is a good example of an only child whose early life was clearly shaped by the intellectual and cultural vitality of the community in which she grew up. "I lived on the west side of Chicago, in the Lawndale district, which was a large area for Jewish immigrants at that time," she recalls. "It was a very intellectual atmosphere. . . . It was not unusual for kids to teach other kids in that neighborhood" (Pless: 1). In particular, when Pless was about thirteen years old, her Sunday school teacher, Samuel Karlin—at that time an undergraduate at Illinois Institute of Technology—was impressed by her intelligence and took her under his wing, tutoring her in Hebrew and calculus. He would later go on to a distinguished career of his own in pure and applied mathematics.[4]

Tilla Weinstein feels that being an only child—and specifically, being *brotherless*—gave her the freedom to escape some of the conventions of femininity. "I was less aware of what it meant to cater my aspirations to the female role," Weinstein says, "than I would have if there had been a brother in the house who was treated differently" (Weinstein: 2). During the crucial years of early adolescence, when the social imperative to be pleasing toward boys can easily derail a young girl's intellectual development, she benefitted from the almost exclusive company of other girls like herself:

My close friends in high school didn't have brothers either. We were at a special girls' junior high. And I think that helped us to plan our lives in ways that weren't as restricted as they otherwise would have been. Also, we were the best hope of our families, whereas if there had been a son, the son would have been the best hope for the family. (Weinstein: 2–3)

Twelve of the women interviewed had just one sibling. It is curious to note that none of the women who had a younger brother has very much to say about him; older brothers, on the other hand, are recalled with warmth and affection. Grace Bates and her older brother were bound together by the experience of losing their mother at an early age and remained extremely close. Her brother completed high school on the eve of the Depression and went directly to work; as soon as he was able, he sent her a little bit of his salary every week to help out with her expenses during high school and college. Similarly, Herta Freitag and her brother, Walter Taussig, separated by less than a year in age, were extremely close as children and have remained devoted lifelong friends. Even Anneli Lax, who characterizes her childhood as having been extremely unhappy, remembers her half-brother, five years older, with fondness. What these three women have in common is that each was drawn closer to her brother through the experience of shared adversity: for Bates, a mother's death; for Freitag and Lax, escape from Nazi persecution.

Five of the women grew up with one older sister. Such two-sister dyads are frequently studies in contrast, particularly if the sisters are close in age, as the relationship between Barbara Beechler and her older sister Virginia clearly illustrates. Margaret Marchand and her older sister show a similar sort of split. While her sister "was not interested" in college, Marchand became increasingly passionate about education as she grew older (Marchand: 2).

Marchand's parents had emigrated from the Ukraine to Canada as adults. "Their education . . . since they were from the peasant class, amounted to approximately three or four grades, with no English," she recalls. Her father was extremely ambitious, studying English in night school and then using his night school textbooks to teach his children. "By the time we went to school" says Marchand, "we were bilingual" in English and Ukrainian. Her father firmly believed "that it was obvious you couldn't get out of poverty without an education" and encouraged his daughters to succeed in school. Marchand embraced her father's attitudes and aspirations. Her sister, by contrast, adopted attitudes much closer to those of her mother, who "was of the school that said women stayed home and got married and had babies" (Marchand: 1–2).

Other two-sister dyads show less tension and less conflict in values. Domina Spencer's older sister Vivian played a key role in her home schooling and introduced her to the wonders of mathematics. The lack of conflict between the two sisters may be attributable in part to the difference in their ages: Vivian, twelve years older, may have been more like an aunt than a sibling in the sense that her presence in the family did not engender heated competition for parental attention. Indeed, for much of Domina Spencer's childhood, she had the best of both worlds: the companionship of a sister, the attentions of an only child. But even when two sisters are close in age, each can carve out a distinctive niche for herself that allows her to express her individuality without conflict in the family unit. Both Evelyn Granville and Joyce Williams had one sister about a year and a half older than herself. In each case, the two sisters each pursued their own interests and goals without engendering opposition within the family.

The remaining fourteen of the women interviewed came from families of three or more children; of these, five were eldest children. It has been argued that eldest children often exhibit greater conformity, while younger children are more rebellious, imaginative, and nonconforming; but everything depends on the values of the family in which one was raised in the first place (Helson 1968; Sulloway 1996).

The experiences of Rebekka Struik, the eldest of three daughters, illustrate this quite clearly. By comparison to the larger society, her own outlook is decidedly original, iconoclastic, and nonconforming. But in a manner of speaking, Struik has indeed been a conformist to the unconventional intellectual and political values of her family. Both of her parents held Ph.D.'s in mathematics and were active throughout their adult lives in left-wing political causes. In religious outlook, she says, "both of them were atheists, but my father comes from a Dutch Calvinist background, and my mother comes from a Jewish background" (Struik: 2). It was assumed in the family that the children would pursue careers in science; and in fact, both Rebekka and the younger of her two sisters, an ecologist, have done so. It is the middle sibling "who rebelled against the overall intellectual atmosphere in our family," devoting herself to marriage and children and becoming actively involved in the Catholic Church (Struik: 2).

Winifred Asprey, Joan Rosenblatt, and Edith Luchins are also eldest children; they, too, can arguably be called conformists to the values of their families. But those families were rich intellectual environments, in which curiosity and creativity were valued. The divergent backgrounds and the intellectual passions of Winifred Asprey's parents, for example, created an intellectually exciting home life. "We were allowed to read anything under the sun," she recalls, "[with] no restrictions whatsoever in the house" (Asprey: 2). Asprey's two younger brothers likewise embraced the predominant values of the family; as adults, one became a chemist, the other, an historian.

"Laterborn" children seem to face a mixed bag of advantages and disadvantages growing up (Sulloway 1996). In large families, younger children often benefit from a great deal of attention, not only from the parents but from the older children, who help to raise them. This situation is best illustrated by the childhood experiences of Mary Dean Clement, who grew up in and around Nashville, Tennessee. Her father was a Methodist minister, while her mother—a graduate of George Peabody Normal

School (now part of Vanderbilt University)—had been a schoolteacher prior to marriage. Clement was the youngest of five children and recalls that "our whole family was always studying. My father always had a book in his hand. . . . And we read voraciously. . . . I was just *extremely* lucky to be born into a family that considered intellectualism fun" (Clement: 4). While the whole family was steeped in education, reading, and learning, her mother and her next older sister had the biggest impact on her early life. She attributes much of her affinity for mathematics to their influence:

[F]rom the way I was brought up, I was quite ignorant of what is common knowledge, to wit: that girls can't do math. Therefore, being ignorant, I liked math and I did it without knowing I wasn't supposed to. Tied in with that is the fact that both my mother and my next older sister majored in math in college, and both of them got me started when I was curious as a child, not by working the problems for me but by showing me how to do them. (Clement: 1)

Clement, like Domina Spencer, received much of her formal education prior to high school at home; her education was a project in which the whole family participated. The Clement household was a vibrant place, the center of life and learning. Prior to high school, Clement reports, "I did not have any friends, real friends," but her family seemed to provide for her all the social life she needed. In those days, she recalls, "I was miserable if I was out of the cocoon of my family" (Clement: 4–5).

But many of the other women who were laterborns did not receive the same quality or degree of attention from their siblings and parents that Clement did. I have already mentioned Jean Walton, who was the fourth of five daughters, and Dorothy Maharam Stone, who was the fifth of six children. Their recollections of childhood do not include memories of being doted on by older siblings, nor do they include stories about intellectual excitement at home. As children, they were both loners, intensely focused on the importance of doing well in school. While both had fathers who valued learning, family life did not have the same quality of intellectual adventure that it had for Mary Dean Clement, Winifred Asprey, or Herta Freitag. In their jobs as rabbi and headmaster, respectively, the fathers of Maharam Stone and Walton created and nurtured intellectual communities in the world outside—but seemed disinclined to do so at home.

For some laterborn children, older siblings helped to mediate and re-solve difficult conflicts with parents. Vivienne Morley's childhood experi-ences provide a case in point. She was the youngest of four daughters, widely spaced apart in age. Her parents, Polish emigrants from impover-ished backgrounds, were diametrically opposite in education and out-look. Her father had very little education and a rather poor command of the English language; her mother, by contrast, excelled at languages, including English, Polish, and Yiddish. Her mother was the dominant parent, valuing education for her daughters, not for its own sake but "as a social asset": her chief concern was "in marrying her daughters off, marrying her daughters off *well*" (Morley: 2). Thus she greeted the aca-demic and professional success of her first three daughters (the oldest became a history teacher, the next an accountant, the third a physician) with considerable dismay.

When Morley began to manifest scholarly, bookish inclinations as an adolescent, her mother sought to undermine her intellectual aspirations, hoping instead to see her youngest daughter attain the kind of conven-tional married life she had wanted so badly for herself. Morley's next older sister—nine years older than she—intervened twice to help relieve the tension, inviting Vivienne to leave her parents' home in Philadelphia to live with her in Oregon. For several years—first in high school and later in college—Vivienne Morley lived thousands of miles away from her mother's interference, enjoying the freedom to explore her academic interests with the support of a sister who nurtured and shared them.

Larger families, by virtue of their sheer size and diversity, can some-times provide emotional and material resources that smaller families can-not. Likewise, the extended family and the larger community can play a role in shaping the future direction of a young girl's life that is almost as significant—in some cases, *more* significant—than that of the nuclear family. The experiences of Alice Schafer provide a case in point.

Unlike the other women in this study, Alice Schafer never knew her parents and never experienced the environment of a nuclear family while growing up. While still an infant, she was placed in the care of two aunts, Pearl and Beulah Dickerson. For the first few years of her life, Pearl Dick-erson, who lived in Richmond, engaged a nurse to care for Schafer. Ultimately, she went to live in rural Scottsburg, Virginia, with her "maiden aunt" Beulah (Schafer: 2).

The Dickerson sisters provided a model of female self-sufficiency that Alice Schafer might not have seen had she grown up in a traditional family. In particular, Pearl Dickerson was able to keep her job during the Depression, despite being married to a man who also worked:

She worked for the federal government and, fortunately, did not lose her job during the Depression, although she was married. She had no children. She was greatly criticized for not giving up her position, since her husband had not lost his. And [to keep her job,] she had to point out how many people she was caring for. What amused me was that she included her maid as one of the persons she was supporting because she did have help seven days a week. (Schafer: 3)

Back in Scottsburg, Beulah lived on a farm and was able to support herself partially by renting the land out to others who worked it; she was also one of the many people who received support from Pearl in Richmond.

While the Dickerson sisters and their family provided Alice Schafer with substantial emotional and material support growing up, she was also raised, in a manner of speaking, by the larger community of Scottsburg. She recalls, for example, that while she was on the high school debating team, some of her best coaching came from the town postmistress. When the time came to consider application to college, a prominent man in Scottsburg offered to help her get a scholarship.

Evelyn Granville is the only African American woman among the interviewees. More than any of the other women interviewed, her childhood was shaped by interactions both within her nuclear family and within the extended family and the larger community of which she was a part. Granville grew up in the African American community of Washington, D.C., where women—even, and perhaps especially, women with children—had always worked for a living. Her mother had graduated from high school and worked as a secretary until she married; when she separated from her husband, she returned to the workforce as a maid. Her mother's sister, who was always very close to the family, "had more of an academic bent" and prepared for a job in teaching. "But the first year she was out," Granville says, "she didn't get an appointment." Ultimately, "both she and my mother ended up working at the Bureau of Engraving and Printing in Washington" (Granville: 1–2).

As a very young child, Granville was keenly aware of what was possible for African American men and women in Washington, D.C.; the experiences of her mother, her father, her aunt, and their extended family and friends illustrated the range of possibilities. She noted, first of all, that the opportunities open to women and to men, limited as they were, were also quite different. Her father worked at "a variety of jobs," selling vegetables from a truck during the Depression and occasionally working as a janitor; but the only work he was able to find in the federal government was as a messenger. Black women, on the other hand, seemed to have somewhat greater employment opportunities at the federal level, working not only at the Bureau of Engraving and Printing but also in the Government Printing Office (Granville: 3).

Beyond any doubt, Granville knew that education was the key to expanded opportunities and a better life:

Black people valued education. Black people—we weren't *black* then, you understand; we were *Negroes* or *colored* then—colored people knew that in order to progress, you needed a college education. And, of course, our greatest, our most evident role models then were black, colored, Negro teachers. And of course, the teachers represented success; they represented stability in the community. They lived better than anybody else, and so naturally, you wanted to be like they were. And in order to be like they were, you needed an education. . . . So it was never a question of whether you were *going* to go to college. (Granville: 3)

In terms of their impact on her early life, Granville's teachers assumed an importance comparable to that of her family. "I saw black women—attractive, well-dressed women—teaching school, and I wanted to be a teacher because that's all I saw. I was not *aware* of any other profession" (Granville: 3).

Thus for Granville—and, indeed, for all of the interviewees—the early development of personality, intellectual interests, and academic ambitions was founded on crucial interactions within the family and in the wider community as well. In the next section, I consider how these childhood experiences helped to create key images of adulthood on which lives and careers are built.

Images of Adulthood and the Formation of Ambition

How does a young woman form the desire to become a mathematician? More generally, how does she form her aspirations? How does she form an image of adult life in general, and how does she conceive of what her own adult life will be like? At what point in the process does the pursuit of mathematics become a part of the plan?

The images of adult womanhood that girls form as they grow up are constructed in part from their observations of the behavior of the adults around them. They observe their mothers and fathers, male and female relatives, men and women in the community and at school. They observe how men and women interact in society and in the larger culture. They often tailor their aspirations so as to conform to the positive aspects of the models of adult womanhood they see around them. But when the images of adult womanhood to which they are exposed seem too oppressive, too rigid or constraining, they can form ambitions that are openly defiant of the cultural norms for women. Indeed, for many young girls, the formation of ambition combines elements of conformity and defiance.

One of the central features of the myth of the mathematical life course is the idea that mathematical talent is innate, emerges early in childhood, and must be nurtured from an early age. For the women in this study, however, mathematics was just one among many interests they had as children. In only comparatively few cases does mathematics appear to have been a chief interest from early childhood on. Mary Dean Clement is the only one of the women interviewed to suggest that her mathematical interests and abilities were already manifest in infancy. She recalls that her three earliest memories, dating back to "before I was three years old," were very clearly of a mathematical nature. "Two of these definitely had to do with geometry, and the third had to do with my disturbance that something was not symmetric. I think these memories and understandings occurred before my family had much opportunity of training me" (Clement: 1). Grace Bates also recalls that she had an interest in mathematical games as a toddler, but for most of the women, an interest in mathematics did not emerge until after they had started school.

Like Evelyn Granville, many of the women in this study decided early on that they wanted to become teachers. Because Granville grew up in a community where adult women were *expected* to work, she came to idolize her teachers because they were her only model for what a successful, productive, well-respected woman of color was like. For many of the other interviewees, teachers were frequently the *only* adult women they had ever met who worked outside the home and maintained an ongoing, vital connection to the world of knowledge and ideas. Thus it was natural for them to decide at an early age in favor of a teaching career; only later did mathematics emerge as the subject they would go on to teach.

For some, a desire to teach grew naturally out of the love of learning. Tilla Weinstein reports, for example, that "no matter which grade I was in—first, second, third—among the things I wanted very much to do was to teach at that level" (Weinstein: 3). For Margaret Marchand, becoming a teacher meant a new life beyond the confines of the rural Manitoba community in which she grew up:

The only thing that I remember is, even as far back as third grade, since the schoolteacher was the only other role model that I saw, outside of the farmers and their families, I was determined that I wasn't going to stay in that area. I was at least going to be a schoolteacher. (Marchand: 2)

But in a few cases, the interest in mathematics and the aspiration to teach emerged nearly simultaneously. Herta Freitag provides the best example of this:

[A]t that time I kept a diary, which I still have. It is, of course, in German. And I was twelve years old, and what the diary says is, "School is sort of all right, but it seems to be the case that getting an education is equated with memorization, which I find simple but boring. But I have finally found a subject where I don't have to memorize: mathematics. I can just think it out." And then comes the next entry: "I like mathematics more and more from day to day, and I now know that I want to become a mathematics teacher." And then, the last entry, still at twelve, I rather like: "I have changed my mind. I do not want to become a mathematics teacher. I want to become a *good* mathematics teacher." (Freitag-1: 1)

"It was my mother, really," says Margaret Martin, "who made the decision that I was going to have a career where I could earn a living, rather than being married and then dependent upon my husband completely" (Martin: 3). Martin's mother had been a teacher but had quit her job before she married. Disappointed by a marriage that was neither particularly happy nor particularly satisfying, her mother was determined that Martin would not make the same mistakes she had. "From the time I was fairly young, it was assumed in the family that I would be a teacher," Martin recalls. "In those days, there were not many careers for women, and my mother regarded teaching as the most desirable" (Martin: 1). Martin accepted her mother's decision without question.

By contrast, Tilla Weinstein *observed* her mother's unhappiness and resolved for herself that she would have an independent life and not allow herself to feel trapped in a marriage:

I might be a lawyer; I might be a clothing designer; I might be a buyer for a department store. I don't think I ever thought in terms of owning a business. Entrepreneurial adventures were not for me! I thought at different times after high school of being a writer. But probably the most persistent sense was that I wanted to teach. Behind all of this—which is important—was the knowledge that I was not going to put myself in the situation which I saw my mother in. I expected to work in order to not be dependent. I didn't see myself as working in order to attain some great status or to excel particularly. I wanted control of my existence. And even though I expected to get married and I expected to have children, I never wanted to have to stay in a bad marriage. (Weinstein: 3)

What Margaret Martin and Tilla Weinstein have in common is that each began to form her vision for the future in reaction to the adverse circumstances of her mother's life.

Martin and Weinstein were certainly not the only women who formed such reactive ambitions. Alice Schafer's earliest aspirations arose in defiant rejection of the limitations she saw placed on girls and women everywhere she turned. In particular, she began to have an *emotional* investment in mathematics when, as an elementary school student, it was first suggested to her that she might not be *able* to do mathematics. When Schafer and one of her female classmates were skipped from the third to the fourth grade in their Scottsburg, Virginia, school, the teacher told Schafer's Aunt Beulah, "Well, I think the girls will do all right in fifth grade, but I don't know whether they will really be able to do long division." Schafer responded indignantly to the teacher's concern: "That was a challenge I couldn't take! I was determined to learn to do long division in the fourth grade. So I think that was my first *feeling* about mathematics" (Schafer: 1).

From an early age, Alice Schafer was keenly aware of the many kinds of discrimination a woman could suffer. She had seen, in particular, how a woman teacher could lose her job for reasons unrelated to her performance: "One of my favorite teachers, Glenna Snead, lost her job when she got married. . . . Glenna was an excellent teacher, and I thought,

how horrible that she could not teach because she had gotten married"
(Schafer: 7). Schafer's Aunt Pearl—undoubtedly among her most powerful, positive female role models growing up—was a woman who worked
and who did *not* lose her job when she married. Outraged by the example
of Glenna Snead and emboldened by the example of her Aunt Pearl, Alice
decided early on that she, too, would be a working woman: "I was just
going to have a career, period" (Schafer: 7). Moreover, she resolved that
her life would be *different* from the average working woman's: she would
have a career, but it would *not* be in teaching.

In the twenties, thirties, and forties, most young women—no matter
how unconventional their career aspirations—nevertheless anticipated
that they would marry and have a family when they grew up. Even as
children, however, many of the women in this study seemed intuitively
aware of the difficulty of combining marriage and motherhood with
meaningful work in the world. For Alice Schafer, the best plan was to get
her intellectual and professional life underway and worry about marriage
later. In fact, as a young girl she had more or less decided on the best
age for marriage: "I didn't know to whom, but I was going to get married
when I was twenty-seven" (Schafer: 7).

But Dorothy Maharam Stone, who saw marriage as yet another among
many obstacles standing in the way of her goals and dreams, simply resolved *never* to marry:

What I thought I would like to do, was to learn. But I didn't see any way I could
get a job. In those days, many jobs were closed to Jews. They couldn't get jobs.
For instance, there were many Jews among my classmates, and no Jewish teachers
at all, any more than there were Jews working for banks or anything like that.
So I thought, "I'm not going to get married. I'm going to have a house all by
myself, rooms all by myself. Then I can do what I want in the evenings, educate
myself." Meanwhile, I suppose I knew I didn't have the sort of personality to sell
fancy clothes in the department store. I thought I might get a job in the domestics
department. That was probably the easiest way for a Jewish girl without any
specific training to get a job. (Maharam Stone: 3)

Vera Pless, by contrast, recalls having no such misgivings about marriage as a child. While she had no clear sense of how far she would pursue
her education, she seemed to know with certainty that she would marry
and have children. "I was going to be a wife and mother," says Pless.
"Even though my mother was professional and everything, she had a big

family here, her sisters. And it was quite clear that all the kids, the women, the girls, were supposed to be wives and mothers" (Pless: 3). Whatever tensions Pless's mother may have experienced between her work as a dentist and her family life may have been lessened—or at least obscured from view—by the fact that her dental office was right at home.

But Rebekka Struik observed firsthand the intense and often painful conflict her mother experienced between her responsibilities as a wife and mother and her desire to do mathematics. Unable to resolve the conflict in her own life, Struik's mother transmitted confusing messages about work and family to her children:

She filled me with conflicting pictures of my possibilities as an adult. . . . It was clear she loved math; she talked about how wonderful it was. [But] there was this whole conflicting message: that you've got to marry and have kids but if you're going to be a scientist, no man you'd want to marry would marry you! And so this was all very disturbing. And I was determined I would not be stuck with just kids and diapers for a sizeable part of my adult life. (Struik: 3–4)

Augusta Schurrer, like many of the women interviewed, developed a dual set of ambitions in childhood. Growing up in poverty in New York City, she was imbued from a very early age with a sense of the importance of work and self-sufficiency, although she wasn't quite sure what she would do to make a living. At the same time, she believed that at some point in the future she would be married and no longer working outside the home: "Somebody would take care of me, after a while" (Schurrer: 4). Clearly envisioning both work and marriage as separate phases of her future life, Schurrer expended little energy worrying about what might happen if these disparate visions ever came into conflict. Like many of the other women interviewed, she skipped grades in school, was sharply focused on her studies, and remained content to live her life day to day.

Indeed, for most of the interviewees, childhood was a time to learn, to observe the world around them, to wonder and to dream. With the approach of college, however, each woman was called on for the first time to make plans—however tentative—for her future. And in the act of making these plans and following them through, she began to experience the excitement and the conflict of life as a woman in mathematics. In resolving the inevitable problems that followed, each woman drew on her own resources—her family, her community, her intelligence, her ingenuity, and the images of adult life she had seen or created in childhood.

4

High School and College

Adolescence: Crisis and Opportunity

A child's basic personality emerges early, in the context of the family in which he or she grows up. But during adolescence both boys and girls begin to form their identity as persons, trying on various roles in a social context in an effort to define themselves in a way that will guide them toward adulthood while at the same time distinguishing them from others. At its best, adolescence is an exciting time of experimentation, growth, and development.

For adolescent boys, the pursuit of individual goals and dreams is normally not incompatible with what the culture expects of them. But for teenage girls, adolescence is a time of crisis, as the inner drive to follow their own interests and inclinations comes into direct conflict with the cultural expectations of marriage and motherhood, which call on them to attend primarily to the needs of others. Perhaps the key developmental task of adolescence, for both girls and boys, is to create an authentic identity within the context of mutually rewarding relationships with others.

The women who became mathematicians in the 1940s and 1950s began to focus their academic interests on mathematics during high school and college. What factors enabled them to develop an interest in mathematics—a field traditionally dominated by men—during these pivotal years of adolescence and young adulthood? First and foremost, they had to have a natural, intrinsic *curiosity* about mathematics. Next, they needed to have the *opportunity* to explore mathematics in some depth and breadth—by taking advanced courses in mathematics or science, for example, or by having the financial means to attend college. Moreover, they needed the support and encouragement of key mentors and role models, both male and female.

In addition to all of these positive factors, quite a few of these young women seemed to thrive on a moderate amount of adversity: direct challenges to their mathematical interests and aspirations, if not overwhelming or preemptive, emboldened them to persevere. Many girls and young women respond to the message that they cannot or should not do mathematics with defiance, as if responding to a dare. For many of the interviewees, such provocation was an important factor motivating them to continue their study of mathematics. Indeed, the sense of specialness associated with being a woman in mathematics was something that a few of them actively enjoyed and that inspired them to go on in the field. At the same time, some of these young women were treated unfairly in high school and college, their mathematical talent ignored or downplayed because of their sex. Although all of them were able to persevere in mathematics despite these disadvantages, how their lives might have unfolded differently in the absence of such bias remains an open question.

In the pages that follow, I examine the complex interplay of internal motivation, social conditioning, encouragement, provocation, discrimination, and opportunity in the high school and college years of the thirty-six interviewees. Their experiences illustrate how young women of their generation navigated the rocky terrain of adolescence and young adulthood to take their first decisive steps toward a career in mathematics.

High School as an End to Isolation

The American high school experience in the twenties, thirties, and forties was quite different from what it has become in the latter decades of the twentieth century. Then as now, high school was a center of sports and social life for young people, but the academic significance of high school was much more widely understood and respected. High school was typically viewed within the family as a serious academic undertaking, a mark of distinction, a privilege, and an important rite of passage into adulthood.

What were the purposes served by high school for the interviewees as young women? For many of them, high school spelled an end to isolation, as they finally met other people—teachers and peers, male and (especially) female—who shared their intellectual interests and outlook. As we have seen, many of the interviewees grew up in intellectually stimulating home environments, where books and ideas were valued and discussed. But for nearly all of them, high school provided the first place where their ideas could be shared, tested, and developed in a social context.

Prior to high school, Mary Dean Clement, Domina Spencer, and Margaret Marchand had undertaken a good deal of their formal education at home. Clement, for example, interacted only infrequently with children her own age during her elementary school years and recalls the tremendous discomfort she felt when she asked a young playmate about her reading preferences: "She looked at me cross-eyed" (Clement: 4). When Clement reached high school age, her parents sent her to Ward-Belmont, a private girls' school in Nashville, which all of her older sisters and one of her nieces had attended before her. The experience of single-sex education proved invaluable to her at this stage in her life. At Ward-Belmont, the education of young women was taken seriously, and there, for the first time, she had girlfriends who shared her love of learning and were her intellectual peers. In fact, all of her closest friends from those days went on to professional careers.

From her earliest childhood, Clement had felt a natural affinity for geometry. A talented geometry teacher at Ward-Belmont helped her to develop her geometric intuitions by wedding them to logic and proof:

The geometry teacher was superlative. I think, probably, in my whole mathematical career, she was *the* best math teacher. She really taught you what geometry was about, what a proof was about, how to follow a logical argument, and how to gain an understanding of what a proof is. And now, if I'm confronted with a simple problem, my instinct is to try to solve it by plane geometry rather than by algebra or calculus or whatever. (Clement: 5)

In general, the high school years were a time of social growth and intellectual maturation for her.

Domina Spencer's transition from home schooling to more formal education was entirely accidental. In the fall of 1933, Domina, her sister Vivian, and her mother headed off to Philadelphia, where Vivian was to begin work toward the Ph.D. at Penn. Domina and her mother had planned to stay in the city for just a week or two, helping Vivian to settle in her new apartment before returning to their home in Oberlin.

While wandering around downtown Philadelphia one afternoon, Domina and her mother happened upon the grounds of the Friends' Select School. The Spencers had recently developed an interest in the Society of Friends and were naturally curious about what a Quaker school might be like. During their impromptu visit, they ran into the headmaster:

And he got to talking to Mother, and he discovered I'd never been to school except . . . one summer session. And he begged her to let me go to Friends' Select School while we were visiting. He said, "Come for six weeks." Well, I went four years! It was entirely unplanned. (Spencer: 8)

At Friends' Select, Domina Spencer gained a broad and deep exposure to art, history, science, and languages. Moreover, she says, "I learned to feel it was very important to be active in politics" (Spencer: 10). She engaged her peers in political debates; she acted in plays; and in general, she gained crucial experience in the give and take of intellectual life in a social setting. Her education at Friends' Select did *not*, however, include mathematics; she continued the private study of mathematics at home with her sister. Mathematics thus retained its special status, as an art to be studied with a private tutor, much as one might undertake the independent study of music or painting.

For Clement and Spencer, home schooling prior to high school was essentially a choice made by their families. The early educational isolation of Margaret Marchand, by contrast, was a product of circumstances. The remote, rural Manitoba community in which she grew up had "a one-room schoolhouse with one teacher and eight grades"; she took grades nine and ten "by correspondence from the department of education in Winnipeg" (Marchand: 3). She attended the one-room school every day, completing her first two years of high school work—including courses in mathematics, physics, and Latin—by independent study. "The lessons went to the teacher and the tests went to the teacher," she recalls. "All she did was give me the test and monitor it and then send it back" to

Winnipeg for grading (Marchand: 3). Although she found herself in the company of other children every day, her work was done in complete isolation, without assistance, direction, or feedback from either the teacher or her fellow students.

Her final two years of high school were completed at a boarding school sixty miles away from home. The move to "the little town of Teulon" liberated her from her solitude; there, for the first time, she found companionship and camaraderie among her peers. Understandably anxious about whether she would be able to perform as well at this new school after years of independent study, her fears were quickly laid to rest: "I ended up being suddenly right up with the rest of them. . . . I was able to do quite well compared to everybody else" (Marchand: 3). This realization did wonders for her confidence: she had, in fact, been able to educate *herself*.

In particular, the principal of the Teulon school, Mr. Robson, who was also the mathematics teacher, recognized that she had a "special aptitude" for mathematics and was the first person to suggest to her that she should go on to the university. "I was a shy little girl from the farm," she says, but her high school teachers recognized her talent and encouraged her to expand her horizons still further (Marchand: 1).

Margaret Marchand's high school experience led her, literally and figuratively, out of her isolation and into contact with a wider world of people and ideas. Two other women also attended boarding high schools, although not out of necessity. Grace Bates attended the Methodist-run, coeducational Cazenovia Seminary in New York State; and while she chafed under the rules and regulations that required boys and girls to be chaperoned at all times, she relished the freedom of being away from home. Patricia Eberlein's parents, concerned about the quality of the public schools of South Bend, Indiana, sent her away for high school to Shipley, a girls' college preparatory school in Bryn Mawr, Pennsylvania.

While none of the other women made the same sort of physical journey that these three did, many of them clearly viewed high school as an important new stage, a step forward into a world of expanded social and intellectual horizons. It is not an exaggeration to say that a significant number of these women saw high school as a place of intellectual adventure.

High School as a Place of Intellectual Adventure

Edith Luchins, the scholar of her Brooklyn, New York, family, resolved to make the transition to high school an adventure in travel as well, within her family's limited means:

I wanted to go out of town to high school, and that's what I did! Instead of going to one of the Brooklyn schools, I went across the river. I took the trolley every morning, with my girlfriends; a bunch of us had decided to go out of town. We went to Seward Park High School on the Lower East Side. We wanted to go out of town! Here was this inexpensive out of town experience. You could go to any high school you wanted to, but no one ever thought of going anywhere but those that were handy and nearby. And that was a very positive experience for me. (Luchins: 7–8)

At Seward Park High School, Luchins excelled in mathematics; there, for the first time, she began to savor her uniqueness as a woman in a "man's field." She tutored her classmates in mathematics, helping at least one struggling student to prepare for the Regents' Exam. At the same time, she explored her other interests and talents; she was elected president of the Seward Park High School's General Organization—"a little bit of the politician in me" (Luchins: 8).

Similarly, Augusta Schurrer was transported from her relatively poor, German-speaking neighborhood in New York City into the company of other bright students and talented teachers at Hunter College High School. A succession of nurturing elementary school teachers, who truly cared for the welfare of their students, encouraged and directed her toward Hunter, where admission was by competitive exam and tuition was free.

At Hunter, Schurrer was introduced to a range of subjects she had not previously imagined. "I loved languages," she recalls. "I loved anything that came from books," including English, history, and physics, although she confesses, "I was a klutz in the lab!" (Schurrer: 4). It was at Hunter that she first set her sights on college. Faculty members from Hunter College came to the high school to talk with the students about their options for the future, and Schurrer credits one of these visiting professors with inspiring her to consider a career in the mathematical sciences:

I came from a poor family. We had to make a living. And at one point, J. Hobart Bushey—who was professor of statistics at Hunter College—came and gave a talk at the high school and said that this was a field in which you could earn a lot of money, it was mathematical, and if you enjoyed mathematics, this would be something to go into. (Schurrer: 1)

It was then, for the first time, that she made the connection between studying mathematics and making a living. This experience caused her to think seriously about setting a course for the future in which mathematics would play a central role.

In her home city of Vienna, Herta Freitag attended not high school but its nearest European equivalent, gymnasium, which lasted for eight years (from age ten to age eighteen) and was entirely single-sex. In gymnasium in the 1920s, Herta Freitag found the presentation of subjects to be thoroughly integrated, "completely noncompartmentalized." In mathematics, for example, algebra, geometry, trigonometry, and analytic geometry were completely interwoven. She likens her gymnasium education to a spiral, which each year progressively "widens and deepens, until you get to the top"—which, in mathematics, was calculus (Freitag-1: 1).

The teachers in gymnasium were often remote and sometimes inscrutable, and they were treated with reverence and respect. In fact, what was most striking and important to Herta Freitag about the culture of the gymnasium was that knowledge itself was revered and respected. She recalls that in Vienna in the twenties, class distinctions were based more on education than on money, so that education was held in the highest esteem.[1] Freitag recalls that, of the twenty-seven students in her gymnasium class,

twenty-six went on to study in university. And we were, at the time, in the culture—very, very interesting, I remember everything clearly—terrifically idealistic and terrifically very animated by the desire to study. For instance, about that person that didn't go on to university, the talk was, "Leave it to that dumbbell! She'll probably get married right away!" (Freitag-1: 6)

Herta Freitag enjoyed a peer culture in gymnasium in which education was valued and those who dropped out of school, even at an advanced level, were looked down on. Moreover, the values of the school were reinforced in the larger culture—a consistency in values that she sees as sadly lacking in the United States.

Even in U.S. high schools, however, these young women were able to find or to create social worlds in which intellectual and cultural values were shared. As in the Viennese gymnasium, this was frequently easier to accomplish in a single-sex environment. But having even one friend who supported her intellectual interests in high school could be enough to keep a young woman going. It was just such a friend that Barbara Beechler had, from seventh grade onward, in her boyfriend, Bob Blair. Beechler and Blair were "the two stars" of their public high school (Beechler: 4). Although they rarely took science or mathematics classes together, Blair was the *one* peer with whom Beechler could not merely share but cultivate a passionate interest in mathematics and science—physics, in particular. It was this companionship that helped her sustain a sense of intellectual adventure during her high school years, years that were made more difficult for her and for her family by World War II and her father's extended hospitalization with tuberculosis.

High School as a Place of Limitations

While high school can be a place where intellectual, cultural, and social horizons expand, it can also be a place where painful limitations are placed on the curious. These limitations often reflect biases against exceptional students—female students, in particular. What is striking among the group of interviewees is that nearly all of them viewed such limitations as challenges to be overcome, as provocative experiences that served to reinforce their resolve to excel.

The limitations of Mary Ellen Rudin's high school were of the equal-opportunity variety, since they affected male and female students alike. Because the school was so small, course offerings were extremely limited. "We had minimal quantities of mathematics and science," Rudin recalls. "I did not have anything like trigonometry, for instance. . . . Everybody took the same thing" (Rudin: 3). These limitations did not, apparently, affect the seriousness with which a bookish student might approach his or her studies. In the remote Texas town where Rudin went to high school, this meant that students simply invested more time and energy in what was available to them.

At Cazenovia Seminary in New York in the 1920s, Grace Bates encountered an unexpected obstacle when she tried to take as much mathematics as was offered at the school. "I was taking the usual elementary algebra and then geometry," she recalls, "and I wanted to go on in my senior year with intermediate algebra. And they said there that I'd have to take a history course [instead]" (Bates: 3). While the school argued that the history class was a requirement, there is no evidence that male students were similarly barred from the algebra course. The fact is that it was highly unusual for a female student to take all of the available mathematics courses at Cazenovia in those days; her exclusion from the algebra class may well have been a case of sex discrimination.

How did Bates respond to this affront? "I squawked—and I wrote my dad," she recalls (Bates: 3). Her father, who was secretary to the commissioner of agriculture in New York state, used his influence to bring pressure to bear on Cazenovia Seminary. Authorities within the state government argued on her behalf that while she could take a history course "any time," it was crucially important for her mathematics education to proceed without a break. Cazenovia was willing to modify its stance under pressure, and Bates was able to take another year of mathematics.

At other high schools, however, the limitations encountered were more overtly sexist. As we have seen in chapter 3, Alice Schafer's earliest emotional investment in mathematics was shaped by a teacher's doubts about her ability. At her small, rural high school in Scottsburg, Virginia, she excelled in mathematics; but her mathematics teacher stood firm in his opposition to the idea of her going on in the subject:

In high school, there were three teachers; of course, not many courses were offered. I liked mathematics and Latin and history best of all the courses. When I was applying to go to the University of Richmond for college, I needed a scholarship. And I said to the principal, who was the one who taught mathematics, "Would you write a recommendation for me?" He asked, "What do you want to major in?" and I told him these three fields. "Oh," he said, "if you want to major in mathematics, I won't write for you because girls can't do mathematics." (Schafer: 1)

Her fury at his statement simply added fuel to her determination to prove him wrong. "I don't remember any more whether he wrote for me or not," she says. "My Latin and history teachers probably wrote for me" (Schafer: 1). She got the scholarship and went on to major in mathematics at the University of Richmond.

Jean Walton recalls receiving subtler but no less powerful messages about the mismatch between girls and mathematics. After attending public elementary school in Bucks County, Pennsylvania, she attended George School, the private Quaker boarding school where her father was headmaster. She enjoyed mathematics and continued taking coursework in it all through high school. "I took an advanced class. In those days, advanced high school math was solid geometry and trigonometry," she recalls. She does not remember anyone objecting to her decision to continue on in math. But she was aware of her unique status as the only girl in the class, and she remembers feeling distinctly out of place there:

I remember very vividly a class session, during that senior year, when the teacher said, "Now, any of you who are going on to college and planning to go on in mathematics, I want to urge you to buy a slide rule and to get familiar with using it." And, as you probably know, in any case, in those days there were no calculators, and the slide rule played a very important role. But the thing that I remember so vividly was the clarity of my conviction that "He's not talking to me. Because girls don't take math! I, of course, won't be doing that when I get to college. I'm doing it now because it's kind of fun, but that won't be what I will do, and I don't need to get a slide rule and I don't need to learn how to use it!" (Walton: 1–3)

Whether the teacher was communicating this to her directly or not, she clearly held the conviction that while it was all right—permissible—for her to be taking mathematics, it was *not* all right for her to be taking it seriously. Mathematics was for fun—it was not something she *needed* to do—although the very next year, she went on to Swarthmore College and majored in the subject:

I never got a slide rule, and I never learned how to use it. And before very long I became embarrassed over not knowing, and I tried to hide the fact that I didn't know. And I was able to do this; I got along perfectly well. I used tables, and I guessed, and I never used a slide rule. (Walton: 3)

During her teens, Walton says, "I was . . . very aware of the fact that men were the people who knew things, and men were the authorities. And I was not a person who had a mind of her own at all at that stage of my life" (Walton: 3). These attitudes, though not *explicitly* communicated, were almost certainly part of the culture at George School, and they were also reinforced at home, by the example of her headmaster-father. Their impact on her was long-lasting, undermining her sense of accomplishment in mathematics for many years to come.

At first glance, Lakeview High School on Chicago's North Side, with an enrollment of some five thousand students, could not have been more different from the high schools that Alice Schafer and Jean Walton attended. Yet Violet Larney, who graduated from Lakeview in 1938, experienced much of the same sort of discrimination there that Schafer and Walton had experienced in far smaller, more intimate, more bucolic high school settings. Larney studied mathematics throughout all four years of high school, earned straight As in all of her courses, and came out at the top of her class of well over a thousand students. Yet her mathematics teacher tried to discourage her from continuing in mathematics, much as Alice Schafer's teacher had done:

When I told my mathematics teacher that I was going to go to college and major in mathematics, even though I was a straight-A student in his mathematics classes, he said, "Well, there isn't much call for mathematics these days. It's the social sciences that are getting all the interest." It was 1937 when he said this to me! And he said, "You might not get a job if you go into mathematics, especially being a woman." (Larney: 3)

Despite her obvious academic talent, Larney's high school teachers never really encouraged her to make plans for her future education. She was an only child whose father died when she was a high school senior, and it was her mother—not a high school teacher or counselor—who insisted that she continue on to college. When it came to choosing a college, Larney decided to go to Illinois State Normal University on the offhand suggestion of a family friend. Larney chalks her experience up to the fact that "they gave very poor advice in high school in those days" (Larney: 3). But at the same time, she is also acutely aware of what became of the male salutatorian of her class, who went on to work on the Manhattan Project at the University of Chicago: "He went farther than I did, even if he did get one B!" (Larney: 1). One has to wonder what sort of advice was being given to *him*.

Violet Larney, who should have been the pride and joy of her teachers, was simply not taken seriously because she was a girl. She responded to the situation in the only way she could: by making do and pressing onward with the resources available to her. But still, sixty years later, she wonders what it would have been like if she had been able to explore her options more fully at the time.

High School as a Place of Encounter with Mathematics

High school is the place where, under the influence of a particularly inspiring teacher, a good many of the women interviewed had their first serious encounter with mathematics. For Margaret Marchand, as we have seen, going on to university and going on in mathematics were two sides of the same coin: neither would happen without the other, and perhaps neither would have happened without the intervention of her teacher, Mr. Robson, at a crucial juncture in her high school education.

Lida Barrett traces the origins of her active interest in mathematics to her participation in the junior high school mathematics team. "Texas had an interscholastic league," she recalls, "and schools had math teams. When the football team went to play football, the math team went and 'played math' against the math people at the other school" (Barrett: 1). In particular, Barrett attributes much of her early facility in math to the training techniques of the coach, Miss Emma Finch, who opened her eyes to a sense of how mathematics *worked*. Possessed of this understanding, she came to regard mathematics as "easy"—as something that she was fully capable of doing, and doing well.

From her adolescence until well into her twenties, Patricia Eberlein had eclectic interests; she would not make the decision to dedicate herself to mathematics until long after graduating from college. But in high school, she found algebra and geometry especially interesting. Before moving off to high school at Shipley in Bryn Mawr, she took a high school algebra class in South Bend, Indiana, where she had an unusual encounter with the mathematician Karl Menger.[2] "Menger, who was at Notre Dame at the time, used to proctor my exams. He even gave me papers he'd written, which of course I couldn't read!" (Eberlein: 2). While it may have been preposterous for Menger to have given Eberlein his research papers to read, his attention and encouragement had a lasting impact on her. Later on, at Shipley, she was one of only two girls at the school who enrolled in a full four years of mathematics. Her encounter with algebra and with

Menger in South Bend laid the groundwork for her much later emergence as a mathematician.

Anneli Lax's educational experiences prior to college were repeatedly disrupted by her family's flight from the Nazis. Despite the seemingly interminable interruptions, Lax had key experiences with mathematics during her high school years that influenced her long thereafter. Her interest in mathematics was awakened by Euclidean geometry, which she first studied in lyceum in Berlin around 1935. The constructions and the logic of geometry afforded her "the perfect sort of escape: I didn't have to look up anything; I didn't have to consult libraries or books. I could just sit there and figure things out" (Lax: 1).

As she was leaving Germany, her teacher from the Berlin lyceum urged her, "Whatever you do, whatever happens to you, make sure you go to university" (Lax: 4). On her arrival in the United States, she took geometry yet again from "a lovely old lady, Miss Eaton," at a high school in Queens (Lax: 3). These two significant encounters with geometry convinced her that logic was what made mathematics satisfying and pleasing. Because she could not see its underlying logical structure, the algebra class that followed on Miss Eaton's geometry course was a disappointment: "It looked like a lot of tricks. It was only much later that I saw that there was a foundation to the stuff. . . . I never liked rules that didn't have a basis that I could understand." Despite her dissatisfaction with algebra, Anneli Lax came out of high school "fascinated with mathematics" (Lax: 4–5).

Jean Rubin and Edith Luchins attended New York City public high schools—Rubin in the forties and Luchins in the thirties.[3] An all-around good student, Jean Rubin enrolled in the college preparatory program at Andrew Jackson High School in Queens. During her junior and senior years there, she developed her particular interest in mathematics. She says, "I had a very interesting instructor. He had a Ph.D. in math, and I think he turned me on. I really became interested in mathematics after that" (Rubin: 1).

By contrast, Edith Luchins had been interested in mathematics from her earliest days in elementary school. But it was in high school that her mathematical experience broadened and deepened, and her confidence grew. She took all the mathematics courses offered by the school; she participated in the math club; she tutored her fellow students; her mathematics teachers occasionally asked her to assist them with grading or to take charge of the class for a time. She grew not only in her love of mathematics but in her desire to teach it as well. For Luchins, as for nearly all the women interviewed, high school set the stage for a much fuller and deeper engagement with mathematics at the college and university level.

The Transition to College

In recent years, college attendance for both women and men has become the norm rather than the exception, but (particularly for women) this is a relatively recent development. In 1930, just over 10 percent of all women eighteen to twenty-four years old attended college; during the years of Depression and war, this proportion continued to increase but had not yet reached 20 percent by 1950 (Solomon 1985: 64). As Herta Freitag has observed, in Austria as in other European countries, class distinctions have long been based on education. In American society, where the class structure is somewhat more complex, there has nevertheless been a strong correlation between educational level and socioeconomic advancement.

For the women who became Ph.D. mathematicians in the forties and fifties, the transition from high school to college was their first decisive step toward a professional career. Unlike the prospective college student of today, most of them applied to just one or two schools for admission; their choice of institution was frequently constrained by the availability of scholarships and the limited financial means of their families. Twenty-two of the women interviewed lived at home for all or most of their college years and made their choice of college based on the assumption that they would do so. For example, growing up in Pittsburgh and coming from a poor family that did not particularly encourage her to go on, Dorothy Maharam Stone realized that she would have to go to a local school. She chose Carnegie Tech (over the University of Pittsburgh) because they offered her a scholarship and because they were willing to let her major in mathematics rather than "home economics or secretarial work or library work" (Maharam Stone: 1).

Anne Lewis Anderson grew up in Lynchburg, Virginia, home of Randolph-Macon Woman's College, which she attended on a scholarship as the top-ranking female student in her high school class. "As far as I know," Anderson recalls, "I had never considered any other college. From the time I was a child, Randolph-Macon was there and I was headed toward it" (Anderson-2: 2). Similarly, Margaret Willerding, Lida Barrett, Jane Cronin Scanlon, and Augusta Schurrer attended hometown colleges because they were affordable and, in some cases, free.

Many of the women were quite young when they started college. Augusta Schurrer, who had been promoted through the "rapid advance system" of the New York City schools, was fifteen when she made the transition from Hunter College High School to Hunter College in 1941. Vera Pless was also fifteen when she enrolled in the "Hutchins College" of the University of Chicago in 1946; she was admitted after two years of high school on the basis of her performance on an entrance examination.[4]

Although they had their share of fun, the women who attended college close to home did not lead the sort of all-encompassing campus life that students at residential colleges and universities typically experience. But Marie Wurster, who grew up in Philadelphia and attended nearby Bryn Mawr College, had the best of both worlds. For her first two years, she lived at home and commuted to Bryn Mawr. After that, she says, "I got a second small scholarship which enabled me to live there the last two years," and she was able to enjoy the experience of living on campus (Wurster: 3).

The students who went away to college or university did so for a variety of reasons. Throughout her childhood in Iowa, Winifred Asprey felt that she was on a trajectory toward Vassar College—much as Anne Anderson felt that she was destined to go to Randolph-Macon. At Vassar, she would be following in her mother's footsteps and continuing a family tradition. Mary Dean Clement decided to attend Wellesley College in fulfillment of a family tradition of a different sort. "My parents had a very strong belief that a person should explore the rest of the world," Clement says. "And we did not have the money or the facilities for just traveling for the sake of traveling. So it was encouraged that all of us go off to school" in a different part of the country. She was attracted to Wellesley because it was a women's college, close to the cultural amenities of Boston, and yet located "in a village that had an identity of its own" (Clement: 6).

Evelyn Granville attended Dunbar High School, the academically oriented high school for black students in Washington, D.C. "Dunbar had a great tradition of sending students to the Ivy League schools," Granville recalls. "Men, boys went to Amherst, Dartmouth, Yale, Harvard, and the women went to the women's colleges." With the encouragement of her high school homeroom teacher, Granville set her sights on Smith College, "even though my mother was certainly not financially able to send me" (Granville: 4). When she was admitted to Smith, her entire family bore the expense:

I did not receive a scholarship the first year [at Smith], and I was told later that they didn't see how in the world as poor a child as I could afford to go there. In my [high school] class, two went to Mount Holyoke, one went to Wellesley, one went to Vassar, one went to Bates. There were several [who went to] Ivy League schools. Their families were professionals, and they probably were able to afford it. In the group that went away to school, I was probably the only one who was financially inadequate. But the first year, my aunt helped my mother. Of course, after the first year, I got scholarships. I lived in a co-op house, worked during the summers, and I was able to [pay for it]. It was not a financial burden after the first year. (Granville: 4–5)

For many young women in the 1940s, going away to college could be an act of rebellion—a conscious effort to *break* with family tradition. This is most evident in Barbara Beechler's decision to attend the University of Iowa in the fall of 1945. The war years had been difficult for the Beechler family. During her father's illness, her mother went to work at the draft board; her older sister lived at home, worked part-time, and attended Augustana College, a Lutheran school in their home town of Rock Island, Illinois. As a senior in high school during 1944–1945, Barbara decided that she needed to get away from home. Without seeking the advice of anyone—neither her parents, nor her teachers, nor her boyfriend, Bob Blair, who was about to go into the military service—she chose the University of Iowa, about fifty miles from home. Why did she choose Iowa, when the University of Illinois, in her own home state, was actually a greater distance away? Perhaps it was the adventure of *leaving the state* where she had spent her entire life to that point—while at the same time having the option of easily coming back home should she feel the need.

News of her admission to the University of Iowa in the spring of 1945 came as a surprise to her parents, who had no idea that she had even applied there. They anticipated that she would go to Augustana College, just as her sister had done. Beechler, who came from a Catholic family, sought the assistance of a higher authority to win them over to her point of view:

I went to the local priest, and enlisted his aid, and got him to tell my parents— ah, shame on me!—that I shouldn't go to Augustana because I would be required to go to chapel. And it was not a Catholic chapel. [Augustana College is] Lutheran, Swedish Lutheran. And of course, this was total nonsense, but at any rate, it worked. (Beechler: 5)

Barbara Beechler's decision to attend Iowa granted her a degree of freedom from her family, while at the same time maintaining family ties. It was rebellious, original, independent—but still relatively safe. For Margaret Marchand, on the other hand, each new phase of her education *necessitated* her traveling farther away from home. Her transition from the family farm, to the high school in Teulon, and on to the University of Manitoba in Winnipeg was a continuous expansion of her intellectual, social, and geographical horizons. Tilla Weinstein's journey from New York City to the University of Michigan in the early 1950s was a blending of both kinds of experiences. She went away to school to get away from the discord between her parents at home; she chose Michigan because there would be relatives living close at hand: "The branch of the family that I was most taken by—the branch of the family that had done well and seemed to enjoy life—lived in Detroit and encouraged my parents to send me to Michigan" (Weinstein: 4).

At the same time, Michigan was for Tilla Weinstein what Manitoba had been for Margaret Marchand—a genuine expansion of her intellectual and cultural experience:

I was going to major in English—there was an English honors program that I loved, and there was a philosophy honors program that I loved. And I had a cousin who had just gotten a Ph.D. at Michigan, was just finishing, and kept twisting my arm and making me stick with calculus. . . . I loved everything about being there. I didn't miss New York because I knew I could go back. It was the most ideal intellectual environment. . . . I had never been in a situation where I could work all the time and get eight hours of sleep at night. I did notice that I didn't see any poor people, I didn't see any sick people. And that was a change from New York. It was a young population. (Weinstein: 4–5)

Whether close to home or far away, college life brought these women significantly closer to the integration of their intellectual and cultural interests with their future lives as adults. And for nearly all of them, college or university was the place where they first made a long-term *commitment* to mathematics.

Choosing Mathematics from a Range of Options

Several of the women interviewed—Clement, Bates, Larney, Marchand, Freitag, Gassner, Hahn, Luchins—knew that they wanted to go on in mathematics even before they graduated from high school. But for most of the others, the conscious choice to go on in mathematics was made during the college years, under a variety of circumstances.

Some of them came to mathematics through an initial interest in physics, shifting their interest to mathematics in a quest for greater clarity, logic, and certainty. Jane Cronin Scanlon made the switch from physics to mathematics after taking a course in quantum mechanics as a junior at Wayne State University. She felt uncomfortable with the material even though she did well in the course: "I felt that I didn't understand anything, [and] I couldn't explain what I didn't understand!" Mathematics seemed like a refreshing change:

One of the things I liked about mathematics was that there were clear statements: "Now, this is what we're going to *assume*. And this is what you have to *prove*." This was clear-cut. Now, a physical science can't be clear-cut in the same way because one doesn't *know* what's going on; one has to take guesses. And when it's a good guess, then it's a great theory; but I, of course, understood nothing of this [at the time]. (Cronin Scanlon: 5)

In short, mathematics offered a clarity that no empirical science could match.

Barbara Beechler actually completed a bachelor's degree in physics at the University of Iowa, but as she moved on into upper-division physics courses, her desire to understand the underlying mathematics made it difficult for her to keep up with the physics. Like Cronin Scanlon, Beechler continued to do fairly well in physics courses but understood less and less of what she was studying. At the same time, she had had some marvelous teachers in mathematics and logic who recognized her talent and took a personal interest in her. Her positive experiences with mathematics ultimately led her to go on in mathematics rather than physics in graduate school.

For quite a few of these women, far more divergent fields competed with mathematics for their attention. Biology was the chief competing interest for both Anne Lewis Anderson and Lida Barrett. Anderson decided in favor of mathematics when she began to tutor her fellow students in mathematics and discovered how much she enjoyed teaching the subject. For Lida Barrett, the path toward mathematics and away from biology was more convoluted. "I had started out in biology but had trouble with my eyes. Microscopes were not binocular, and I wasn't using my eyes well together," Barrett recalls. "The eye doctor wouldn't let me continue with biology, so I switched to math for a year" (Barrett: 2). During her year-long hiatus from biology, she took sophomore calculus with a graduate assistant whose inexperience translated into extremely poor teaching. She was "disgusted" by the experience and had all but decided to switch back to biology when Hubert Bray, head of the mathematics department at Rice, intervened. Bray did what he could to persuade her to stay on in mathematics, and Barrett realized that, despite the poor teaching, it was mathematics that had captured her imagination.

Several of the women—Asprey, Wurster, Granville, Rudin, Lax—had dual interests in both languages and mathematics. Marie Wurster, who entered Bryn Mawr in 1936 at the height of the Depression, quickly decided that "mathematics was a more practical field to major in" (Wurster: 2). Similarly, Evelyn Granville says, "When I got to college, I realized that there's more to French than the language. There's French literature and all that, and I decided I'd better stick with mathematics" (Granville: 4). Mary Ellen Rudin, on the other hand, managed to complete a dual major in mathematics and Spanish, although going on in Spanish was an option she seems never to have seriously contemplated.

Anneli Lax entered Adelphi College, then a small women's college on Long Island, in the late 1930s. "I was torn between languages and mathematics," she says. "I liked languages and I was good at them. On the other hand, I was fascinated with mathematics." Although mathematics was much more of a struggle for her, she continued to derive pleasure from the fact that she could do mathematics just by "thinking about it by myself but not studying the books" (Lax: 5). In the end she chose to major in mathematics, despite—or perhaps because of—the fact that mathematics was more challenging.

Winifred Asprey entered Vassar intending to major in French and stuck it out for quite a while. She rapidly discovered that "everybody in class, it seemed to me, had had a French governess or lived in France," while her own background in French was relatively "spotty" by comparison. To succeed in French classes, she memorized entire sections of the textbook, saving them for the appropriate moment in class. "I waited until the discussion swung anywhere near what I had prepared, and I flung my arm into the air," Asprey recalls, "and when [the teacher, Mademoiselle Monnier] called on me, I rattled forth exactly what I had memorized. So it sounded as if I were speaking at rat-a-tat speed" (Asprey: 7–8).

This deception proved successful in all of the classes she took with Mademoiselle Monnier. But deep down Asprey realized that she really wasn't learning the French. As a scholarship student, she was ineligible to spend her junior year in France, which would have been her first real opportunity to develop her language skills in a French-speaking environment. Reluctantly, she set out in search of a new major.

Mathematics provided an attractive alternative. She had been taking mathematics electives all along, and doing well in these courses required much less of an effort than was required for success in French:

By this time, mathematics was *really* getting interesting. We were up into calculus. [Math] was my easiest subject. It was the one place that I did not have to worry. . . . It was the subject that I found that if I did all the rest of my homework, I'd say to myself, "Now, you can spend the rest of your time on math!" And so, it suddenly dawned on me that it might be something to major in. Well, I began majoring in it, really, my junior year. (Asprey: 9)

In contrast to Anneli Lax, who was intrigued by the challenge and the unfamiliarity of mathematics, Asprey chose mathematics over French because it was easier, more familiar, and more natural to her.

Tilla Weinstein's path to the mathematics major is perhaps the most unusual of any in the group. She entered the University of Michigan in 1951 as a major in English and philosophy, and she remembers her coursework as a source of intense pleasure and delight. "I *read* for the first time in my life, extensively. I'd never been a reader," she says. "I *wrote* seriously for the first time in my life and enjoyed that enormously." She was enrolled in the honors program in both fields and uses words like "wonderful" and "splendid" to describe the programs and courses in both subjects (Weinstein: 6).

Like Winifred Asprey, she had continued taking mathematics as an elective during her first two years of college. As a freshman, she took an honors calculus course with the geometer Hans Samelson. "I worked very hard," Weinstein recalls. "I got an A. But I felt I didn't understand anything, and in retrospect, I *didn't* understand very much! I suppose the only thing that shows I was *potentially* a math student is that I *understood* that I didn't understand" (Weinstein: 7).

After an uninspiring third semester in calculus, her cousin, a doctoral student at Michigan, suggested that she try a course in foundations of mathematics taught by R.L. Wilder:

It was *perfect* for me: you needed no knowledge. He threw out an axiom, gave you a few models for the axiom, gave you a few definitions involving undefined terms, suggested one or two things that could be proved, and let us loose. The whole content of the course, when you looked back, was on a stack of index cards: axiom, definition, theorem. I enjoyed it enormously. It was self-contained and beautiful. . . . You didn't learn many facts. You didn't learn many skills. But you came away with a sense of how things actually came together and how very *much* you get out of very *little*. That was the first time I had enjoyed math since I was in high school. And I was willing to take the plunge into advanced calculus. (Weinstein: 7–8)

Wilder's course set the stage for the unexpected turn of events that followed. At the end of her freshman year, she received a proposal of marriage from a young man from back home in New York. "In those days, if someone you loved, loved you, you got married," says Weinstein; and that's exactly what she did, "June after my sophomore year" at Michigan. Marriage meant that she would leave Michigan and transfer to a college in New York, where her new husband was employed. Her intention was to go to City College because it was free, but she discovered that the requirements at City College would have forced her to retake courses she had already had at Michigan. "It seemed a waste of my college time," Weinstein said, "so I went to NYU" (Weinstein: 4, 6).

At NYU, she discovered to her disappointment that the English and philosophy departments were decidedly less appealing than they had been at Michigan. This might have been reason enough for her to change majors, but there was yet another reason for doing so:

I was marrying someone who intended to go on and get a Ph.D. in English. And it was not comfortable for me to be in the situation of being in the same field as my husband. There was the discomfort of possibly competing; wives didn't compete with their husbands in those days. (Weinstein: 6)

Desperately seeking a major at NYU, it was the memory of Wilder's course at Michigan that brought her into mathematics:

And I remember asking, "What's good here?" And the adviser said, "Oh, there's mathematics." And he never got farther on the list because I had just had the wonderful course with Wilder at Michigan, which I truly enjoyed. (Weinstein: 7)

And she stayed in mathematics at NYU, right on to the Ph.D.

What these stories clearly illustrate is that, for a great many of the women mathematicians of this generation, the decision to pursue mathematics seriously at the undergraduate level was by no means a foregone conclusion. Contrary to the myth of the mathematical life course, these women came to the subject comparatively late, choosing it from among a range of other subjects in which they had demonstrated interest and ability.

Gender Barriers

From elementary school on through high school, college, and beyond, girls and women have often been told that they cannot or should not do mathematics and science. Among the women interviewed for this study, it was at the college or university level that these messages became most explicit, to the point that they became barriers to their educational advancement.

In the thirties and forties, college advisers were frequently unable to take a woman's ambitions in mathematics or science seriously. When Barbara Beechler arrived at the University of Iowa in 1945, not only was she the only undergraduate woman majoring in physics, but she was (so far as she knew) the only undergraduate woman who had *ever* majored in physics! During her first (and last) encounter with an academic adviser in the physics department, she was told that physics was "a very good field for a woman" because "physics studies color, and women really need to know a lot about color, so that they have proper decoration of homes." Beechler says, "I never went near the man for another piece of advice" (Beechler: 7).

Similarly, at Illinois State Normal University, Violet Larney found the advising no better than what she'd had in her Chicago high school. "If you were to major in mathematics and minor in a science like physics," Larney was told, "you'd never get a job. They only want men for that, so you'd better get into secretarial science and business as your minors" (Larney: 3). Larney was aiming for a career in high school teaching. In those days, *men* were routinely hired for jobs in "science and coaching, or mathematics and science," but the message was clear: "a *woman* wasn't supposed to be teaching mathematics and science" (Larney: 3–4). In the end, Larney chose minors in accounting, law, and secretarial sciences; she still regrets the fact that she never studied science seriously at the college level.

As we have seen, from early childhood onward Alice Schafer was acutely aware of the many disadvantages of being a woman. Already in high school, the principal's opposition to the idea that she might go on to study mathematics in college posed a difficult—though ultimately not insurmountable—barrier to her continuing on to study mathematics at the University of Richmond. This was not the last such barrier she was to encounter.

When Schafer arrived at Westhampton College, the women's coordinate college of the University of Richmond, in 1932, she was undecided about her major. The decision was a toss-up between history, Latin, and mathematics; it would be difficult to imagine three more widely disparate fields. At Westhampton, Schafer recalls, "I had a wonderful teacher in history—a University of Chicago Ph.D., as a matter of fact, in ancient history. And it was clear to me that I did not write well enough to be an historian. So that knocked out history." Latin was ruled out when Westhampton's Latin teacher left the school to teach in China; "so that left mathematics" (Schafer: 4).

Westhampton College, like many other women's coordinate colleges, had very limited offerings in mathematics. In particular, "there was one woman who taught mathematics, Isabel Harris. At that time," Schafer recalls, "the women were on one side of the lake and the men on the other; and there were no classes together until you got to the advanced mathematics courses, and the same with the sciences" (Schafer: 4). As it happened, Schafer was the only woman mathematics major in her class. After the sophomore year, she began to take classes on the men's side of the lake and was often the only woman in a class of men.

The head of the mathematics department at the University of Richmond was so firmly convinced that women could not and should not succeed in mathematics that, apparently, he did everything in his power to ensure the outcome:

In my junior year of college, the dean of women called me into her office and said that the head of the mathematics department was flunking all the women mathematics students, for he wanted no women majors in mathematics. The dean told me, "You are to stand your ground." I was too naive for it to have occurred to me *not* to stand my ground!

The course the department head was teaching must have been an advanced calculus course. A woman in my class who was a chemistry major had not been flunked—as she told me later—but had been given a D. I believe that earlier she *had* been thinking about being a mathematics major. Our examinations were not returned to us, and we were never allowed to see them. But a member of the history department told me that the mathematics department head had said in a faculty meeting that I had made 100 on my final examination, and she was shocked when I said I knew nothing about this. (Schafer: 4–5)

Although Schafer does not report what her final grade in the course was, it was clear that her academic performance was so good that the department head was compelled to acknowledge it, at first to his fellow faculty members, and then—grudgingly—to Alice. During her junior year, she won the Crump Prize, awarded for the best performance by a junior mathematics major on a competitive examination:

The head of the mathematics department never congratulated me when he was told that I had won. What he said was, "I never thought *you* would win the prize." I was so angry for so many years that it took me some time to realize that he and the others who graded the papers must have graded them fairly!

By the end of her senior year, the department head "came around slightly," offering to help her get a fellowship to graduate school in mathematics (Schafer: 5).

Among the factors that influenced Alice Schafer to continue on to graduate study in mathematics was her observation of how badly Isabel Harris was treated at the University of Richmond:

Isabel Harris—Miss Harris, as we called her—had never gotten a Ph.D. In fact, I asked her later why she had not; and she, in essence, said, "Why bother? They would never have let me teach anything higher than analytic geometry." "They," of course, referred to the men in the mathematics department. . . . No woman was allowed to [teach the men]. (Schafer: 5–6)

Harris had done graduate work in mathematics at Columbia and Chicago, but she was apparently convinced that completing a doctorate would not improve her situation at Richmond. As a tribute to Harris and as a matter of honor, Schafer resolved that *she* would complete a Ph.D. in mathematics herself. From elementary school on through college, Alice Schafer's determination to study mathematics was fueled again and again by the dual forces of interest and outrage. Her natural affinity for the subject, combined with her sense of injustice at all the barriers to women's success in mathematics she had seen and experienced, served only to intensify her own determination to continue in mathematics to the highest level of which she was capable.

At about the same time that Alice Schafer was struggling against the powers that be at the University of Richmond, Grace Bates was experiencing the limitations of study at the coordinate college associated to Middlebury College in Vermont, where she enrolled as a freshman in 1931. At Middlebury, as at Westhampton, there was just one mathematics teacher on the women's side, Ellen Wiley. Unlike Isabel Harris, however, Wiley had a somewhat freer rein to teach a wider range of courses. On the other hand, at Middlebury, women were not expected to *want* to take the upper-level mathematics classes that were available only on the men's side; women certainly did not take them as a matter of course. As a result, Bates reports, "I had to petition the trustees at Middlebury to get into a course in differential equations my senior year because it was only taught in the men's side" (Bates: 6). Once enrolled in the men's courses, however, she does not report having sensed any hostility to her presence there. At Middlebury, the chief barriers to a woman's seeking advanced education in mathematics seemed to be primarily administrative, while at Westhampton the barriers were more of a personal nature. Both Bates and Schafer were successful in overcoming both kinds of barriers, but they did so in a mathematical climate that was intimidating and inhospitable to them.

A decade later, Rebekka Struik chose mathematics over chemistry as her major at Swarthmore College because there the *chemistry* department was actively inhospitable to women. When Struik entered Swarthmore in the fall of 1945, she was enamored of chemistry, having had an excellent high school course. She did well in first-year college chemistry, but as might be expected in any major, she began to struggle in her sophomore year. Organic chemistry was especially difficult—"like memorizing the telephone book!"—and she turned to the teacher for help. He was anything but supportive:

The chemistry professor was openly antiwoman. This was probably 1947, so there was nothing illegal about it, but by now the chemistry department has cleaned up its act at Swarthmore. But [then,] his attitude was that it was a waste of resources to train a woman in chemistry because "you're just going to get married and have children. And you *should* get married, and you *should* have children." Well, I was doing poorly in the organic, and I went to talk to him, and he thought it would be a good idea if I changed majors. (Struik: 1)

She is convinced that the chemistry professor tried to drive her out of the department, not because of her performance but because she was female. Struik knows of at least one male classmate who did as badly as she in organic chemistry but went on to a distinguished career as a chemist and wonders what the organic chemistry teacher had to say to *him*. It seems unlikely that the male student was given a lecture about the waste of resources that his perseverance in chemistry would entail. "It's one thing to discourage a student because he or she is doing poorly," says Struik. "It's another thing to discourage them because they're of the wrong gender, or the wrong race, or the wrong religion, or whatever" (Struik: 1).

Mathematics was a natural second choice for her after chemistry. Organic chemistry relied heavily on memorization; but in mathematics, "you could just figure it all out. And that's what was important. And that, definitely, appealed to me" (Struik: 6). Moreover, the warm welcome she received in the mathematics department helped to seal the decision. After her unhappy meeting with the organic chemistry professor, "I went in tears to the chair of the math department, Dr. Arnold Dresden, who was a friend of my father's," Struik recalls.[5] "He welcomed me into mathematics, and I've been there ever since!" (Struik: 1).

Rebekka Struik was just one among many women of this generation who were told that an education in science would be wasted on a woman. Domina Spencer was told the same thing a decade earlier when, at the age of fifteen, she sought admission to MIT. Following her junior year at Friends' Select School in Philadelphia, she attended summer school at

MIT at the suggestion of her sister. During that summer in Cambridge, she enrolled in freshman physics and took up sailing, which was to become a lifelong avocation. It was her love of sailing that inspired her, at summer's end, to ask the dean about the possibility of admission to MIT the following year. MIT was suffering from depressed enrollments at the time, and Dean Petré was quick to capitalize on her academic qualifications, guaranteeing her admission on the spot. "They had a nice regional scholarship for the Philadelphia area," Spencer recalls; Petré encouraged her to compete for it, and she won. But when the time came for MIT to invest money in Domina Spencer's education, they were far less forthcoming:

> The MIT tuition at that time was five hundred dollars a year. And you know what they gave me for a scholarship? Fifty! Ten percent of tuition! Ten percent of the tuition is what they gave me, fifty dollars for the year. And I asked them why I got so little. I was third in my class at Friends' Select; there were a boy and a girl [with] grades slightly better than mine, but I was third out of a class of thirty-five, thirty-seven, something like that. And I was sixteen, and they were probably a couple of years older than I was. But they gave me a token scholarship.

Their explanation was simple: MIT told her point-blank that "women are a very bad investment! They just come here to get husbands. They don't do anything with their careers, so why should we waste scholarship money on them?" (Spencer: 12)

Spencer was outraged. In fact, she was *keenly* aware of the value of her education and of what an important investment in her future it really was. She started work toward a bachelor's degree in physics at MIT in the fall of 1937. "I didn't want to do the same thing as my sister," she recalls, "I wanted to *use* mathematics in science rather than studying pure mathematics" (Spencer: 15). Because her MIT scholarship was so paltry, she remained heavily dependent on the financial resources of her family and particularly of her sister, who by this time had completed her Ph.D. at Penn and was working for the federal government in Washington.

Intensely motivated to complete her undergraduate work as quickly and cheaply as possible, Domina Spencer struck a bargain with H.B. Phillips, then head of the mathematics department at MIT. Phillips agreed that if she could pass his course in advanced calculus, he would exempt her from the mathematics classes required for the physics major, a challenge that she easily met. "That's how I got through MIT in two years rather than in four!" Spencer says. "I also went to summer school and took some advanced standing examinations. . . . You just studied the course for a week, took an examination, and it didn't cost a cent" (Spencer: 14–15).

When all was said and done, Domina Spencer was a student at MIT for a total of five years, earning a bachelor's degree in physics and a master's and Ph.D. in mathematics. During her years at MIT, she says, "I kept an eye on the boy that had gotten 'my' scholarship, and he was still fooling around as an undergraduate when I was getting my doctor's degree!" (Spencer: 12). Nevertheless, MIT never admitted their mistake in not offering her a proper scholarship.

Perhaps the most egregious case of gender discrimination at the undergraduate level was that experienced by Tilla Weinstein during her senior year (1954–1955) at NYU. "I became pregnant at the beginning of the summer, quite accidentally," she recalls. When she returned in the fall to plan her schedule for senior year, she was in for a rude surprise:

I wanted to arrange my schedule, sixteen credits in the fall and fourteen credits in the spring, so that I'd have a lighter load the semester I was having the baby, hoping that I could still keep my small scholarship which required fifteen credits per semester. [When I discussed this with a dean,] he informed me that there was no way they would allow me to register as a student in the second semester. As a matter of fact, first he leaned back and laughed, heartily, for about five minutes. What was funny was that I thought I was going to go on with my education even though I was having a baby. He said, "My daughter just had a baby, and she's been in bed for five months," wiping away the tears from his eyes because he thought it was so funny. (Weinstein: 8–9)

The dean told her that to attend classes while pregnant would have been a violation of state law, "that I would be perpetrating a fraud by registering for the class, because I could not *come* to class." She says she has no idea whether what he said was true or not, but she was unwilling to challenge him because "he was also a senior professor in the English department" where her husband was working toward a Ph.D. "To challenge him would have been to incur the animosity of someone who had power over my husband." Ultimately, a compromise was struck:

There was a negotiation between the people at the Courant Institute—probably Lipman Bers—and the dean. I was not allowed to take a full program. I was not allowed to register for a course. I was allowed to register for two reading courses. And, in fact, I continued taking the two graduate courses I had planned to take—linear algebra and complex variables—but on my transcript, they read as reading courses. And then—which was much harder for me than it would have been otherwise—I had to go to summer school to finish up my degree, after the baby was born. (Weinstein: 9)

Grace Bates, Violet Larney, Alice Schafer, Rebekka Struik, Domina Spencer, and Tilla Weinstein were all victims of sexism. But in the thirties, forties, and fifties, the discriminatory remarks and practices they encountered reflected the mores of the times and were easily explained away: "Women don't do science"; "women will just get married"; "pregnant women should not appear in public"; "pregnancy makes women helpless and debilitated"; "pregnancy and education are incompatible." There were no laws, no authorities to whom these young women could appeal for redress of unfair, outrageous treatment.

It is important to remember that these are the stories of those women who succeeded; they had the patience, the stamina, the financial and moral support of families who helped them to overcome the obstacles. And, in many cases, they had the encouragement of teachers and mentors who believed in their abilities and were strong advocates for their personal and professional success.

The Influence of Teachers and the Power of Mentorship

Teachers rarely know the impact that they have on their students. At their worst, teachers can be disorganized, obscure, and discouraging; but at their best, they can convey an enthusiasm for their subject and an interest in people that can capture a student's imagination and fire her ambition. The support, encouragement, and influence of a caring and concerned mentor can help to ameliorate, and even to overcome, the most adverse circumstances, as Lipman Bers's willingness to help negotiate a solution to Tilla Weinstein's pregnancy "problem" bears out. Nearly all of the women interviewed mention that they had at least one teacher at the college level who sparked their interest in mathematics and made an investment in their present and future success. In some cases, they attribute their determination to go on in mathematics to the influence of just one special teacher.

Particularly at the women's colleges, mentorship was often provided by women of the older generation of pre-1940 Ph.D.'s. At Randolph-Macon Woman's College, Anne Lewis Anderson benefitted from the guidance and interest of two Chicago alumnae, Gillie Larew (Ph.D., Chicago, 1916) and Evelyn Wiggin (Ph.D., Chicago, 1936), who took an active interest in her work and encouraged her to go on for the doctorate at Chicago, as they had done.[6] At Vassar, Winifred Asprey had many of her mathematics classes with Mary Evelyn Wells (Ph.D., Chicago, 1915) and Grace Hopper (Ph.D., Yale, 1934)—"the most inspirational people you can possibly imagine!" (Asprey: 9). In particular, Miss Wells was "the greatest teacher I ever had"—the teacher whose style she copied when she returned to Vassar as a member of the faculty (Asprey: 21).

At Bryn Mawr in the late 1930s, Marie Wurster took nearly all of her undergraduate mathematics with Anna Pell Wheeler and Marguerite Lehr (Ph.D., Bryn Mawr, 1925), who were the two permanent members of the faculty during those years. Wurster was particularly fond of Lehr—a student of Charlotte Angas Scott—whom she characterizes as "an outstanding teacher, whom everyone liked and admired" (Wurster: 4). At Smith, Evelyn Granville studied with Anne O'Neill (Ph.D., Radcliffe, 1942), who had been Winifred Asprey's Vassar classmate; and Susan Rambo (Ph.D., Michigan, 1920) who, though "very stern, caused us to be very disciplined mathematicians" (Granville: 5).

Augusta Schurrer recalls that, of her many memorable teachers at Hunter College in the forties, it was a woman, Anna Marie Whelan (Ph.D., Johns Hopkins, 1923), who turned her on to complex analysis, the field in which she ultimately did her doctoral research. Schurrer remembers Whelan's one-semester course as the one "where the fires lit up. . . . I'd never seen anything as interesting in my whole life. I really, really loved complex" (Schurrer: 6). The geometric flavor of complex analysis brought mathematics alive for her in a way that one-dimensional calculus had not.

The few women mathematics Ph.D.'s working outside of women's colleges served as powerful mentors and role models, too. At the University of Minnesota in the 1930s, Margaret Martin was particularly impressed by Gladys Gibbens (Ph.D., Chicago, 1920) and Sally Elizabeth Carlson (Ph.D., Minnesota, 1924), "whom I liked and whom I was interested in because they were women" (Martin: 5).[7] Cathleen Morawetz, whose father was the distinguished mathematician J.L. Synge, says that she "became a mathematician by default" under his influence (Morawetz: 1). But while working toward her bachelor's degree at the University of Toronto in the 1940s, Morawetz recalls that Cecilia Krieger (Ph.D., Toronto, 1930) was the first, and perhaps the only, member of the faculty to take an active interest in her mathematical future.[8]

As an undergraduate at Toronto, Morawetz was committed to an intense course of study in mathematics, physics, and chemistry. But the narrow focus of her studies was a frequent source of discontent. Seeking a change of pace and of scenery, she took a year off between her junior and senior years of study to do war work in Quebec. On her return, still feeling somewhat dissatisfied with her studies, she ran into Cecilia Krieger on campus:

She asked me what I was going to do the following year. And I had been cooking up the idea that I was going to respond to an appeal for people to go and teach school in India. I thought that would be nice and exotic, and I had enjoyed very much going to a different place—Quebec—and I thought it would be fun. So I told her so, and she was horrified. She said, "You're not going to go to graduate school?" (Morawetz: 5)

Morawetz, who was doing well in mathematics but feeling ambivalent about it, protested that she couldn't afford graduate school. Krieger, who was chair of the committee for the Junior Fellowship of the Canadian University Women's Club, pointed out that money needn't be an obstacle and suggested that Morawetz apply for the award. Morawetz followed her advice and won the fellowship; Krieger's determined encouragement had caused her to take herself and her mathematics more seriously.

It cannot be emphasized too strongly that the support and encouragement of male teachers at the college level was critically important for the vast majority of these women. At Swarthmore in the 1940s, Arnold Dresden was an outstanding teacher and a welcoming presence in the mathematics department for both Anne Whitney Calloway and Rebekka Struik. Edith Luchins recalls a whole host of magnificent teachers from Brooklyn College in the 1940s, most of them male: Samuel Borofsky, Walter Prenowitz, Moses Richardson, James Singer, Margaret Young Woodbridge, and—perhaps most important of all—the historian of

mathematics, Carl Boyer.[9] Lida Barrett's brief undergraduate career at
Rice (from June 1943 through February 1946) was filled with memorable
teachers, who "were interested in teaching—and knew what the students
were learning." In addition to Hubert Bray, who persuaded her to stay
on in mathematics, she recalls Szolem Mandelbrojt, who "just insisted
that we all memorize the definition of the limit. His idea was that once
you got the definition straight, you could work on it" (Barrett: 10).[10]

Many other women mention key male teachers from their undergraduate days: Margaret Martin fondly recalls William Hart, head of the department, and Dunham Jackson from Minnesota in the 1930s; Joyce
Williams remembers Warren Loud at Minnesota in the 1940s; Evelyn
Granville remembers Neal McCoy at Smith.[11] However, none of the interviewees had more powerful experiences of male mentorship at the undergraduate level than did Mary Ellen Rudin, Barbara Beechler, and Tilla
Weinstein.

Mary Ellen Rudin's mathematics education at the University of Texas
was almost entirely under the control of one man: R.L. Moore.[12] As she
was registering for classes on her first day at the university, she met
Moore, who promptly recruited her into his mathematics class. "I was
in a class with him every year," she recalls, "from the time I entered the
University of Texas in 1941 until I got my Ph.D. in 1949." His influence
"was the thing that pushed me toward mathematics" (Rudin: 1).

R.L. Moore's idiosyncratic style of teaching has come to be known as
the Moore method. In every course, at every level, he compelled his students to build the subject "from the ground up":

His way of teaching was to present you with things that had not yet been proved,
and with all kinds of things which might turn out to have a counterexample, and
sometimes unsolved problems—that is, unsolved by *anyone,* not only unsolved
by *you.* So you had some idea of what it meant to be a mathematician—more
than the average undergraduate has today. (Rudin: 4)

Rudin has always been supremely self-confident, a quality that was
nurtured by her family in childhood. "Moore played on that," she recalls.
"He encouraged people to believe in themselves as mathematicians because he felt that this was one of the principal tools for doing mathematics—to have confidence" (Rudin: 21–22). In other words, Moore's style
meshed perfectly with Mary Ellen Rudin's talent and temperament.[13]

Moreover, her success in Moore's classes further enhanced her confidence. "I was aware," says Rudin, "that I was *the* person in the class, in almost all cases, who could work the problems" (Rudin: 5). She is firm in her conviction that her association with Moore and her early commitment to mathematics are inextricably intertwined: "I probably would not be a mathematician if I had not worked with Moore" (Rudin: 22).

As an undergraduate at the University of Iowa, Barbara Beechler's mathematical experience was shaped by the teaching of E.W. Chittenden and the advice of Gustav Bergmann.[14] Beechler's first meeting with Chittenden, like Rudin's with Moore, was accidental. As a freshman at Iowa, she was placed in an algebra and trigonometry class that essentially replicated her high school course. After the first exam, Chittenden—realizing that she was capable of doing much more challenging work—invited her to take his course on foundations of mathematics instead. "There were only three people in the class," Beechler recalls, one of them a graduate student in philosophy. "I *loved* this course. I mean, we did things like plot the sine curve geometrically, and I learned about the four-color problem, and I learned all these neat things" (Beechler: 9). At the same time, Chittenden was a disorganized teacher, and his lecturing skills left his students exasperated. But his enthusiasm for his subject was infectious, and his students learned more mathematics in the process of trying to fill the gaps in his presentation than they ever would have learned had he delivered polished lectures. Chittenden maintained an interest in Beechler's mathematical development from the time of their first meeting until quite late in his life.

After he completed a stint in the Navy, Bob Blair joined Barbara Beechler at the University of Iowa; he became a physics major, and once again they resumed taking classes together. At Blair's suggestion, she began taking classes with the philosopher Gustav Bergmann, who introduced them to mathematical logic. Beechler's knowledge of both mathematics and German made her the star of the class:

I remember reading Gilbert Ackerman's *Mathematical Logic;* it was in German, it hadn't been translated. Gus was wonderful; he was sweet to me. I can remember, there were two rather inept philosophy graduate students in that course, and they didn't know any mathematics and they didn't know any science, and they really weren't catching on to this. And the mathematical logic was very algebraic and formulaic, and I loved it. And I can remember Gus leaning over the table—he was a very ugly man, with buck teeth and, you know, spittle coming out all the time—but he would lean over the table and he would say, "All right, Barbara darling, tell the little boys what it says." *[Laughs]* 'Cause those poor guys couldn't read German, either! (Beechler: 14)

Barbara Beechler regards Gustav Bergmann as "a very important influence. . . . I was stimulated beyond anything I'd ever done before. I loved the logic, I loved the formality. I had never really heard anyone as bright as he. He spent time *with* us and *on* us" (Beechler: 15). But his influence was a mixed blessing. When, as a senior at Iowa, Beechler realized how disillusioned with physics she had become, Bergmann dissuaded her from any thought of pursuing graduate work in philosophy by telling her, "It is no field for a woman." His suggestion was that she continue in mathematics and, specifically, that she work with Chittenden: "You and Chittenden will have a wonderful time together!" (Beechler: 1). Bergmann's clearly sexist advice came as a great disappointment to her, yet it was not entirely off the mark. In the aftermath of this conversation, Beechler went on to perhaps the most enjoyable research experience of her life—working on a master's degree with Chittenden.

Tilla Weinstein recalls that when she was an undergraduate at NYU in the fifties, she was the only woman majoring in mathematics at the time. Her advanced calculus teacher, Jean van Heijenoort, recognized her mathematical talent and invited her to join NYU's Putnam examination team.[15] It was van Heijenoort who suggested that she should begin taking graduate courses at Courant Institute during her senior year. Weinstein dismisses van Heijenoort's attention to her, insisting that there were two other undergraduates at NYU at the time who were far more mathematically gifted than she.[16] But although she doubted his assessment of her ability, she *did* take his advice. "The next September, I found myself in Lipman Bers's office, asking if I could take his complex variables class" (Weinstein: 8). Bers was to become her most important advocate and mentor.[17]

At their very first meeting, Weinstein told Bers of her pregnancy. "He was tremendously happy and delighted," she recalls. "It was as if I had told a relative" (Weinstein: 8). Not long afterward, it was Bers who intervened on Weinstein's behalf, negotiating the compromise that made it possible for her to finish her undergraduate degree without undue delay.

Thus did Tilla Weinstein begin a lifelong association—with Lipman Bers and with advanced mathematics.

Extracurricular Activity: Dating, Marriage, and Life beyond School

For most of the women interviewed, education was the most important activity of their teenage years. In fact, just over one-third of the women interviewed completed college during the year of their twentieth birthday—or earlier. Education was an all-absorbing preoccupation, akin to a full-time job. But school did not entirely prevent them from pursuing other interests and activities. Plays and concerts; student theatrical productions, college newspaper writing and editing, and student government; and part-time jobs all provided a sense of connection to a wider world beyond studying and abstract thinking.

In many important respects, adolescent women of the thirties, forties, and fifties grew up more slowly than they do now. There were fewer pressures and fewer opportunities for dating and involvement with the opposite sex. On the whole, the lives of these young women were much more bound up with those of their parents and siblings than they were with the formation of romantic relationships.

Even so, dating and relations with young men were very much on the minds of these women, particularly during the college years. For women at coeducational colleges and universities, it was their first real opportunity to interact socially with young men. As is often the case for adolescent women nowadays, these first forays into dating could be awkward and disappointing. Looking back on her social life at Swarthmore in the late 1930s, Jean Walton recalls her involvement with college athletics— hockey, basketball, and tennis—and her many "wonderful female friends," with whom she got along splendidly. "It was a very successful college career, except that it was not successful in terms of the wonderful happy times with boys that you're meant to have in college," she says with some regret. "I had several male friends who loved to get me to help them with calculus!" (Walton: 5).

As it was almost unheard of for young men and women to carry on extensive romantic or sexual relationships outside of marriage, it is not surprising to find a few early engagements and marriages in this group

of women. In addition to Tilla Weinstein, both Patricia Eberlein and Anneli Lax married while in college. Weinstein's first marriage lasted fourteen years, but for Eberlein and Lax these first marriages were rather
short-lived. Cathleen Morawetz, Joyce Williams, Barbara Beechler, and
Edith Luchins all entered on early engagements during their college years;
for all but Beechler, the ensuing marriages were extremely long-lasting.

The success of these early engagements and marriages depended in
no small part on the intellectual compatibility of the partners. Tilla
Weinstein and her first husband were bound together by similarities of
upbringing (they had grown up in the same New York neighborhood),
by shared academic aspirations, and by the early birth of children. Anneli
Lax did not enjoy the same sort of compatibility with her first husband.
"He was quite a bit older than I, and I think he had the feeling that he
should take care of me," Lax recalls. While she tried to share in her husband's interests, working occasionally in his automotive repair business,
Lax feels that he did not make an effort to take part in things that mat-
tered to her:

What I couldn't stand in the long run was that he didn't seem to enjoy my friends.
I had friends and classmates come over to our house, and . . . he would have
very little to do with them. And I felt a resentment there. That didn't work. And
then I just felt that we were growing apart, and I just left. (Lax: 9–10)

Lax's increasing commitment to mathematics left little common ground
on which to continue the marriage.

Perhaps the most unusual of all the early pairings was that between
Edith Luchins (then Edith Hirsch) and the man she ultimately married.
During the summer between her junior and senior years of high school,
Edith enrolled in a WPA-sponsored course in psychology taught by Abraham Luchins, seven and a half years her senior, who was at that time
working toward a Ph.D. in educational psychology under Max Wertheimer at the New School for Social Research and NYU.[18] From the first,
Abraham Luchins was pleased by her interest in mathematics, while Edith
became intrigued by cognitive psychology, especially as it applied to
mathematics education. Over a period of several years, they became in-
separable companions. From very early on in their friendship, Edith and
Abraham Luchins had begun to plan their lives together, and Edith's education in mathematics was always part of the plan. In particular, Luchins

recalls that her husband-to-be "insisted that we not be married until I graduated from college. He wanted to make sure . . . that I had a bachelor's degree" (Luchins: 13).

For many of the other interviewees, there simply were not many opportunities for dating during the college years. Joan Rosenblatt attended Barnard College—the women's coordinate college of Columbia University—while living at home during World War II. With some amusement, Rosenblatt recalls: "We had a little song composed by one of my classmates: 'There Is No Pursuit after Girls Who Commute'! The only men around, practically, were the people in the V-12 program . . . and a midshipmen's program at Columbia" (Rosenblatt: 11). Dating was a rare occurrence during wartime.

For many of the interviewees, however, the joy and excitement of learning at the college and university level in the company of like-minded friends was in itself immensely rewarding and satisfying. Herta Freitag, who spent nearly six years in pursuit of the Magisterium rerum Naturalium degree in mathematics and physics at the University of Vienna, recalls a vibrant and happy social life with many classmates, male and female. But marriage was decidedly *not* on their minds. Their view, as she recalls it in those days, was this: "It's all right to have happy little outings together, but no responsibility, no marriage. That's for old people" (Freitag-1: 17).

For most of these women, undergraduate study had offered them their first serious taste of advanced mathematics. For those few women who were engaged or married, their personal lives had taken on a new structure or were on the brink of doing so. But for all of them, the question became pressing: "Where shall I go from here with my life?" For all of these women, whether sooner or later, the answer seemed to be: "To graduate school in mathematics."

5

Graduate School and the Pursuit of the Ph.D.

Graduate School: Research Apprenticeship and Rite of Passage

What is the purpose of graduate study in mathematics? A college mathematics major normally takes a broad range of courses in the liberal arts and sciences in addition to a generous helping of courses in mathematics. But a graduate student concentrates almost exclusively on mathematics. The master's degree, awarded after a year or two of study, certifies that its holder has mastered a substantial chunk of the mathematical literature at a reasonably advanced level. The Ph.D. carries the student one step further, requiring that he or she undertake significant, original research in mathematics, presented in the form of the doctoral dissertation. First and foremost, then, the Ph.D. certifies that its holder has completed an apprenticeship in mathematical research.

In practice, however, the Ph.D. degree in mathematics also serves as a teaching credential for college and university mathematics faculty and, more generally, as a union card for full membership in the mathematical community. How is it possible to justify using the Ph.D.—which is, after all, a certification in *research*—as a qualification for college *teaching?* This question cuts to the heart of the long and vigorous public debate on the interrelationship between teaching and research (Richardson 1989; Herstein 1969; Boyer 1990). The experience of participating in the creation of new mathematical knowledge through research and writing at the doctoral level provides a prospective teacher of mathematics with an intimate understanding of how discoveries are made and communicated in mathematics—in short, of the way mathematics *works.* Thus, it is argued, even if a Ph.D. recipient never engages in mathematical research *again,* the research experience gained in graduate school is valuable, even indispensable, to his or her future as a teacher of postsecondary mathematics.

Indeed, graduate school in mathematics is not merely a research apprenticeship but also a rite of passage into the mathematical community. In addition to the necessary coursework and dissertation research, graduate students in mathematics are often engaged in the teaching of undergraduates, in preparation for their future as college teachers. Furthermore, they participate in the common life of their department through seminars and public examinations, as well as more informal social events such as teas, parties, and so forth. The social dimension of graduate study prepares the doctoral candidate for his or her future in the communal life of academic mathematics.

In this chapter I explore the various paths by which the women who earned Ph.D.'s in mathematics in the 1940s and 1950s made their way to their educational goals. Contrary to the myth of the mathematical life course, the path from bachelor's degree to Ph.D. is not always clear and smooth; and for the interviewees, the path was complicated by the changing cultural expectations placed on women and by the changing climate in the graduate schools and the mathematical community as a whole. I examine the graduate school experience for the women Ph.D.'s of the forties and fifties, with a view toward understanding whether and how it prepared them for future membership in the mathematical community. What was the social environment of graduate school like? Were the women taken as seriously as their male classmates? Were they afforded the same opportunities to explore their research capabilities as the men? What, indeed, did they expect from graduate school—and did their experiences meet their needs and expectations?

The Ph.D.'s of the 1940s

In the twenties and thirties, when the Ph.D.'s of the 1940s were growing up, the norm for women was to finish school and perhaps work for a few years but then to retire from public life into marriage and the raising of children. Even so, a small but significant minority of women could be found in well-defined and highly visible professional roles. These women were usually unmarried, and the vast majority of them were teachers. Dedicated to the academic life, they could be found in elementary and secondary schools, at a few coeducational colleges and universities, and especially in the women's colleges. Their position in society commanded respect and admiration, and they served as role models for young girls with an academic bent.

There was, moreover, a clearly established career path for a woman planning a future in college teaching: on completion of the bachelor's degree, she might teach for a few years at the secondary level, but ultimately she would enter graduate school and work for a Ph.D. On completion of the doctorate, she might reasonably expect to go on to teach at a women's college, finding a job through an old-girl network of former teachers, dissertation advisers, and friends.

Naturally enough, many of the women who earned their Ph.D.'s in the 1940s aspired to careers in teaching from a fairly early age, though some did not decide that it was *college* teaching that interested them until they had already taught for several years at the high school level. At the same time, many women in this generation of Ph.D.'s made the decision to go on to graduate school *without* a specific professional goal in mind: they loved school, they loved mathematics, and they wanted to continue studying. Nearly all of the interviewees of the 1940s generation of Ph.D.'s *expected* to marry and have children at some point in their lives. The actual marriage rate in this group of women is just under 60 percent: ten out of the seventeen women eventually married, and *all* of them completed the Ph.D. prior to marriage. While five of the women met future husbands while still in graduate school, in no case did a romantic attachment impede a woman's progress toward the doctorate.

The interviewees who earned Ph.D.'s in the 1940s fall naturally into three categories. The first group comprises those fortunate women who attended college during the Depression years of the 1930s and proceeded directly to graduate school before the United States entered World War II. The women in this group are Maharam Stone, Spencer, Anderson, and Martin; among them, the mean elapsed time between receipt of the bachelor's degree and receipt of the Ph.D. is four years. While Anderson and Martin specifically aspired to careers in teaching, Maharam Stone and Spencer had strong *scholarly* ambitions from the outset. It is interesting to note that marriage occurred rather late in this group: while Maharam Stone married just two years after receiving the Ph.D., Spencer and Anderson waited nearly *two decades* before marrying, and Martin never married at all.

The second major category consists of those who earned their undergraduate degrees prior to World War II, whose progress toward the Ph.D. was either postponed or protracted over several years' time. The interviewees in this category are Clement, Schafer, Bell, Asprey, Bates, Wurster, Willerding, Walton, and Steinberg, and the mean elapsed time from bachelor's degree to doctorate in this group is just over eight years. All of these women had significant high school teaching experience, and a few actually taught at the college level, before earning the Ph.D. While some of them were attracted to teaching as a profession, others turned to teaching as the only livelihood that would make full use of their talents and education. Only three of the nine women in this group (Bell, Schafer, Steinberg) eventually married; the reasons for this low marriage rate are unclear. Their younger adult years coincided with the Depression and World War II, when marriage rates were low.[1] At the same time, the women in this group were strongly oriented to work and study and may have sensed that their personal goals were incompatible with the prevailing cultural model of marriage.

The third and final category of 1940s Ph.D.'s comprises those women who earned their undergraduate degrees during the war and proceeded with only minor delays to graduate school. The interviewees in this group are Calloway, Cronin Scanlon, Granville, and Rudin; all four earned their Ph.D.'s in 1949, and their mean elapsed time from bachelor's to Ph.D. is about five and a half years. These women share several important characteristics: first, none of them seemed particularly committed to the idea of a career in teaching; second, all of them began graduate work at or near the end of the war and thus benefitted from the ready availability of assistantships and fellowships, made possible by war-induced shortages of graduate student labor.[2] Finally, all four of them eventually married, two of them more than once.

The sections that follow are devoted to a careful exploration of the graduate school experiences of several of the women Ph.D.'s of the 1940s.

Their stories provide a glimpse of what graduate student life was like and illustrate some of the subtle (and not so subtle) differences in experience between male and female graduate students of this generation.

Single-Minded Scholars: Dorothy Maharam Stone and Domina Spencer

The stories told by Dorothy Maharam Stone and Domina Spencer, two of the earliest Ph.D.'s among the interviewees, provide a glimpse into what graduate study in the late thirties and early forties could be like for those few women who aspired from an early age to do creative work in the mathematical sciences.

Their graduate school stories share a number of common features. Neither woman had teaching responsibilities during her graduate school years; neither woman has much to say about her fellow graduate students; and each maintained a singular focus on her own intellectual interests and goals, moving from bachelor's degree to Ph.D. in just three years. At the same time, their experiences also provide a study in contrasts. Maharam Stone seems to have been a true loner in graduate school, doing research work largely in isolation and essentially without help from mentors. Spencer, on the other hand, had a great deal of interaction with the faculty, holding several of them in high esteem; but in the final analysis she, too, completed her dissertation independently.

Dorothy Maharam Stone worked toward the Ph.D. at Bryn Mawr from 1937 to 1940 on a scholarship to which no teaching duties were attached. She went to Bryn Mawr on the recommendation of her department chair at Carnegie Tech, L.L. Dines. Bryn Mawr, which had been among the preeminent centers of doctoral education for women since its opening in the 1880s, was an especially exciting place to be in the mid-1930s. Bryn Mawr's department head, Anna Pell Wheeler, was instrumental in bringing Emmy Noether—one of the most distinguished of the German refugees and the most distinguished woman mathematician in history up to that time—to Bryn Mawr from Germany in 1933. Until Noether's untimely death in 1935, Bryn Mawr was an especially attractive center of research activity.

When Maharam Stone arrived in 1937, she was caught up in the intellectual excitement of Bryn Mawr:

I mean, this place was exceptional. That's where Emmy Noether came from Germany. And a lot of bright young men—*very* bright young men—were there when they were young. I mean, for instance, Nathan Jacobson, the algebraist, left there to go to Yale. And similarly, various other people [were there for a time]. The chairman was Anna Pell Wheeler, who was the first woman to give a colloquium to the AMS at the May meetings. She did integral equations. Anyway, when I was there, there was Mrs. Wheeler; and also Gustav Hedlund, who also moved on to Yale and Marguerite Lehr, who did not do any research. And usually, or very often, there would be some bright young person who'd be there just for a short time. And they were still very much under the spell of having had Emmy Noether there. There were still stories going around about her. (Maharam Stone: 5–6)

During her first two years at Bryn Mawr, Maharam Stone took several courses with Wheeler and had already begun work on a Ph.D. with her when Wheeler became quite ill. Maharam Stone recalls the feeling that she was left very much on her own. During her final year at Bryn Mawr, she was buoyed up by the arrival of John Oxtoby, who had come to replace Hedlund:

My last year, I had a brilliant course in measure theory from John Oxtoby. But I was at the same time already starting to write my Ph.D., after only two years of graduate training, and not so much of that. . . . And it never occurred to me, in my inexperience—because Bryn Mawr was a very small and informal place, and I didn't know how things were run—I could, if I'd known, have gone to John Oxtoby for help and advice because I was writing in his line. And he was the kind of person—I'm sure he would have been delighted. And also, he wouldn't have had anyone else to talk to; I might have been someone to talk with! But anyway, he didn't see my thesis till it was finished. . . . I regret very much that I didn't have enough *push* to just go talk with him about my problems! (Maharam Stone: 6–7)

Thus it came to pass that during Dorothy Maharam Stone's three years at Bryn Mawr, she worked quite independently of Wheeler and, indeed, of anyone else. Maharam Stone characterizes her Ph.D. as "a homemade product" (Maharam Stone: 7). Neither Oxtoby nor Wheeler saw her dissertation until it was essentially complete.

Although Wheeler did not retire until 1948, Maharam Stone was her last student. Bryn Mawr's preeminence in the graduate education of women was in decline, and Maharam Stone's sense of isolation at Bryn

Mawr may be symptomatic of the changing climate there. But by the same token, Maharam Stone seemed determined to complete her dissertation on her own, and that determination, combined with her own reticence, conspired to keep her from interacting with Oxtoby, who might have been a natural mentor and friend.[3]

By contrast to the situation at Bryn Mawr, the mathematics department at MIT was on the rise when Domina Spencer began her graduate study there in 1939. In his reminiscences of mathematics at MIT, Dirk J. Struik, who was to become Spencer's thesis adviser, writes:

Karl T. Compton, who became president [of MIT] in 1930 . . . understood fully that a modern engineering school can only be first grade if it is also a leading school of science, which at the time meant mainly chemistry, physics and mathematics. Thus began the transformation of MIT from a still essentially undergraduate college into a research institute of the first rank, but also maintaining or improving its educational facilities. . . . Things were moving, and so was the mathematics department. (Struik 1989: 173–174)

Thus Spencer undertook graduate study in mathematics at MIT at a particularly auspicious time.

Spencer's bachelor's degree was in physics, and her original intention had been to pursue a doctorate in that same field. In fact, she had secured a spot in the graduate program in physics at Johns Hopkins for the fall of 1939. But shortly before she was to leave for Baltimore, she had second thoughts about leaving MIT, where she enjoyed sailing and had begun a promising courtship with a young man. She decided at the last minute to stay on at MIT for a master's in mathematics, telling herself that "I needed to learn more mathematics in order to do the kind of physics I wanted to do" (Spencer: 15). While she was able to secure only limited financial assistance from MIT, her family—specifically, her sister—provided the financial support she needed for graduate study. As it happened, the relationship that led her to remain at MIT did not endure, although her interest in both sailing and mathematics *did*. The MIT mathematics department eventually came through with scholarships, and she never earned another degree in physics.

During her first semester as a graduate student at MIT, she pressured the head of the mathematics department, H.B. Phillips, to let her take a course with the legendary Norbert Wiener. Phillips dissuaded her, urging her to take a course in tensors—then called *absolute differential calculus*—with Struik instead. Spencer recalls:

Professor Struik gave the most beautiful lectures that I have ever heard in my entire life. And he did something else that no professor that I'd had in all my undergraduate courses at MIT ever did. He suggested that there were things in mathematics that I, little girl, could do, that were brand new. (Spencer: 16)

Struik had few of his colleagues' prejudices regarding women and mathematics; his wife was a Ph.D. mathematician, and his oldest daughter Rebekka would eventually become one as well. Struik engaged Spencer immediately in reading Eduard Study's *Geometrie der Dynamen,* a book available only in German, which he had long wanted to understand more completely:

I brought home Study, and I got a German dictionary, and I went to work. And that was the end of my being called a physicist—almost, not quite! Struik was so wonderful, the second semester I took two courses from him. . . . I found tensor calculus was the most beautiful subject I'd ever heard of. And I did my master's thesis with him, finding out what Study had said. And then Struik wanted me to find out how to interpret [Study's work] in terms of tensors. And there was a doctor's thesis! So I had to stay in the mathematics department. (Spencer: 17)

Once it had become clear that she was going to stay on in mathematics, she was offered a half scholarship; since women were not permitted to teach in the mathematics department at MIT, she was ineligible for a teaching assistantship. Financial considerations had caused her to complete her bachelor's degree at high speed; finances once again led her to race quickly to the doctorate. In fact, she asserts, her Ph.D. research was essentially complete by the end of her second summer in graduate school. It was at about that same time that MIT gave her a full scholarship for the first time. With her family at last freed from the burden of financing her education, Spencer made the most of her full scholarship during the academic year 1941–42. Her academic adviser tried to dissuade her:

And my adviser, who was Ted [W.T.] Martin—he was supposed to supervise what courses I took—told me, "You don't need any more courses; you just finish your thesis and take your general exams." And I told him, "You think you've finally given me a full scholarship and am I going to get nothing for this except getting to see Professor Struik for an hour once in a while? I'm taking courses!"

And against his advice—he tried his best to stop me—I took a full load of courses that year. (Spencer: 19)

Fiercely independent, chafing against unnecessary rules and regulations, Spencer took what nowadays would be called a proactive stance toward her own education, viewing the MIT faculty as a resource to be used to her own ends. As the United States had not yet entered the war, the stars of MIT's mathematics, physics, and engineering departments were all in residence—and Spencer managed to study with quite a few of them. Though she never did take a class with Norbert Wiener, she took Struik's course in tensors a total of three times. She took relativity with Manuel Vallarta and an electrical engineering course on the workings of Vannevar Bush's differential analyzer. She did not hesitate to take courses outside the area of her dissertation research: she enrolled in a topology course with Hassler Whitney at Harvard (although her dissertation committee compelled her to drop the course on a technicality) and went over to the MIT electrical engineering department to take a course in illuminating engineering with Parry Moon. This last choice had far-reaching consequences, as it marked the beginning of a professional and personal partnership with Moon that would last until his death in 1988.

Although their personal styles could not be more different, the singular, intense focus on research that Maharam Stone and Spencer achieved at Bryn Mawr and MIT is a common thread that binds their experiences together. They came to graduate school not simply to earn a Ph.D. but to become scholars. So strong was their inner motivation that they were able to overcome obstacles they encountered: Maharam Stone persevered despite her sense of isolation; Spencer remained resolute in the face of those at MIT who seemed incapable of taking a woman seriously.

But this intense inner drive is somewhat atypical of the women who earned their Ph.D.'s in the early 1940s, many of whom came to graduate school seeking credentials for college teaching, with little more than an inkling of what advanced mathematical scholarship was like. The experiences of women at the University of Chicago provide a picture of what doctoral study could be like for those women who sought the Ph.D. as the next step in their professional development.

Limited Horizons: Chicago in the 1940s

From 1940 through 1946, six women earned Ph.D.'s in mathematics at Chicago: Katharine Hazard (1940), Mary Dean Clement (1941), Alice Schafer (1942), Anne Lewis Anderson (1943), Janet McDonald (1943), and Marie Wurster (1946). A seventh, Florence Dorfman Jacobson, began work toward the Ph.D. during this period but never received the degree. While there were many other women graduate students at Chicago during this same time period, these seven are apparently the only ones who actually made significant progress toward the Ph.D. Four of them (Clement, Schafer, Anderson, and Wurster) are among the interviewees, and their accounts, together with information from other sources, provide a coherent picture of the graduate school experience for women seeking the doctorate at Chicago in the late thirties and early forties.[4]

All the women interviewed report that the social environment for graduate students at Chicago during this period was friendly and supportive; Anne Lewis Anderson, who was a graduate student at Chicago from 1940 through 1943 and stayed on for two years afterward as an instructor, recalls:

The relations between the faculty and the graduate students, and the relations among the graduate students, were just marvelous. The older graduate students would take the young ones under their wing, you know, and help them out in rough spots, and the professors were, almost without exception, just very warm and helpful. Mr. and Mrs. [W.T.] Reid used to have teas frequently on Sunday afternoons for the graduate students, and the graduate student group would often take one of the professors out to lunch, this sort of thing. It was a very congenial, happy group. (Anderson-1: 5)

Although few, if any, of the Chicago women came to graduate school with specific research ambitions, all of them found dissertation advisers without much difficulty. Hazard and Anderson worked in calculus of variations with Magnus Hestenes; Wurster worked in the same field but with Lawrence Graves. Clement, Schafer, and McDonald wrote dissertations in geometry with E.P. Lane, who was head of the department from 1941 through 1946. Jacobson nearly completed a dissertation in abstract algebra with A. Adrian Albert. All seven—Jacobson included—made fairly rapid progress toward the Ph.D.[5]

In his brief history of mathematics at the University of Chicago, Saunders MacLane says that in the thirties and forties, "women were *encouraged* to study for the Ph.D. degree at Chicago," but they "were not really *expected* to do any substantial research after graduation; the doctorate was *it,* and in many cases the thesis topic was chosen to suit" (MacLane 1989: 145; emphasis added). In other words, women were welcomed to doctoral study, they were taken seriously as future college teachers, but their dissertations were a means to an end rather than an apprenticeship for a future career in research.

The relationship between E.P. Lane and his students is a case in point. Over the course of his career he had a great many women students. He has been described as "a very kind and courteous Southern gentleman" (Clement: 13); indeed, all of his female doctoral students in the 1940s hailed from the South.[6] Perhaps the main disadvantage to working with Lane was that his line of research—projective differential geometry— was highly idiosyncratic and unfashionable.[7] In fact, it seems likely that Lane was aware of the limitations of his research techniques—so much so that he did not hesitate to refer his better *male* students to other advisers:

In 1939, George Whitehead, one of the graduate students, asked Professor Lane for a thesis topic in projective differential geometry. Instead of giving him a topic, Lane gave Whitehead the good advice to work in the newer field of topology with Steenrod; Whitehead later (at MIT) became a leader in this field. (MacLane 1989: 142).

Alice Schafer—attuned since childhood to injustices and slights against women—was keenly aware of the differential treatment of male and female graduate students at Chicago. As a senior in college, Schafer had set her sights on the University of Chicago, where many of her favorite Westhampton College teachers had gone. Her failure to win a scholarship left her "brokenhearted" and unable to attend. "But I didn't give up on Chicago," she says. Instead, she taught high school mathematics in Virginia from 1936 through 1939 to earn the money that would enable her to go (Schafer: 8). She made her way into graduate school at Chicago by attending summer school there in 1938; the impression she made on the faculty—Lane in particular—helped her to win a scholarship for full-time study.

When Schafer arrived on campus in the fall of 1939, she worked on a master's thesis with Lane, who had taken a special interest in her. As she progressed in her graduate coursework, however, she became increasingly interested in abstract algebra and wanted to pursue her Ph.D. with Albert instead. But Schafer felt committed to working with Lane, "and I didn't know how to get out of it," she recalls. "If I had been more sophisticated or more mature, I would have known how to handle it" (Schafer: 14).

Naturally, Schafer was angry when she learned how readily Lane passed Whitehead on to Steenrod. Moreover, she believed that neither Lane nor Albert took women's research potential seriously. "Albert said once, when Florence Dorfman and I were having coffee with him in the coffee shop, 'You two are going to get married.' He didn't quite say, 'Forget the mathematics,'" Schafer recalls, "but I assume that that's more or less what he was trying to say" (Schafer: 15).

It is curious to note that the other female graduate students at Chicago did not seem to feel that they were treated inappropriately. Alice Schafer—who dated a Chicago classmate, Richard Schafer, and married him once they finished their Ph.D.'s—may have been all the more keenly aware of discrimination because of her connection to Dick. The one place where the gender imbalance became clear to everyone, however, was at afternoon tea. At Chicago, as in many other mathematics departments, faculty and students would gather in the department common room for a social hour with tea and other refreshments, usually in advance of a departmental seminar or a colloquium featuring a distinguished visiting mathematician. Then as now, afternoon tea was an important time for students to make contact with professors, to meet interesting visitors, and in general, to discuss mathematics in an informal setting.[8] Alice Schafer recalls that while the male graduate students were free to mix and mingle, "the women were in charge of the teas: ordering the cakes; preparing the tea; washing the dishes afterward. The male graduate students were never asked to do anything" (Schafer: 13). The protocol at teatime sent a clear message to the women students about their future role in the mathematical community: they were to act as facilitators and helpers, leaving the real business of mathematics to the men.

This sort of differential treatment was certainly consonant with the prevailing cultural assumptions about the role of women in public life. Women graduate students were expected to complete their degrees, teach at small colleges, marry perhaps, but never really engage in serious scholarship or rise to positions of authority in the mathematical community. Because these expectations were not fundamentally in conflict with the goals that brought them to graduate school to begin with, these women were not inclined to challenge the system—to *insist* that they be taken more seriously. The net effect was that their graduate school experiences, challenging and rewarding though they were, were somewhat circumscribed. In effect, the institutional assumption became a self-fulfilling prophecy.

Discovering Research: Grace Bates and Jane Cronin Scanlon

As the experience of Grace Bates at the University of Illinois from 1944 to 1946 clearly shows, however, it *was* possible for a woman with curiosity and drive to come to graduate school without specific research ambitions and develop genuine passion for research in the process. Bates, like so many of the Chicago students, came to Illinois intending to write a dissertation in geometry. But when she arrived, she found herself captivated—as Alice Schafer had been—by abstract algebra. Unlike Alice Schafer, she decided to switch fields and to work with a formidable adviser, Reinhold Baer.[9] Baer "sounded like he thought American students were all lazy, stupid, and whatnot. But he didn't act that way, and I really enjoyed his class very much. So I switched my interest from geometry to algebra, and I was able to get him to be my adviser" (Bates: 13).

At the outset, Bates "dreaded" the process of writing a dissertation:

I didn't think I could do it, and I thought it would be all drudgery. Well, it wasn't that way at all with Baer. He gave me the definition of a loop, of which I'd never heard before, and an idea of using a graphic approach to some of the theory he had believed ought to be true, and said, "Oh, go ahead now, let's see what you can do." And I'd fumble around, and I didn't think I had much. I'd go into the study room we graduate students had, and he'd be in the very next day to see. And after I'd give him something that I'd done, he'd come in and say, "Miss

Beets!" He never did get my name right. And he'd say, "Well, this is all wrong. This theorem is all wrong. I don't think it's true." And he used to—not always—make that derogatory a statement. And I thought, "I'm going to stay up all night till I find a counterexample!" And I really worked like a dog, and I got into my study place there the next day, and sure enough he came in and said, "Ah, Miss Beets, I found a counterexample!" And I said, "So did I!" *[Laughs]* And that was the first time, really, that I began to have confidence in myself. (Bates: 13–14)

In spite of her initial anxiety about research—and despite Baer's gruff and intimidating manner—Grace Bates rose to the occasion. In meeting the challenge of working with Baer, she gained the confidence and skills of an independent scholar.

There is a discernible difference between the experiences of the women who earned their Ph.D.'s in the early to middle 1940s and those who earned their degrees in the later years of the decade. This later group of women experienced the early impact of the postwar expansion of mathematics and, specifically, the accelerated pace with which world-class research claimed top billing in the graduate departments of mathematics in the United States. While differential treatment according to sex was still significantly in evidence, women students could be drawn in as much as men by the fervor for research.

The most vivid picture of the postwar graduate research environment is the one drawn by Jane Cronin Scanlon as she describes the mathematics department at Michigan in the late 1940s. Early in her graduate program, she took several courses with R.L. Wilder, a master of the Moore method in which one begins with a handful of definitions and axioms and builds a subject up, as it were, from scratch. In his courses, Cronin Scanlon learned how "to sort out what's axiom and what's proof" (Cronin Scanlon: 9). Early on, she also took an abstract algebra class with Warren Ambrose (Ph.D., Illinois, 1939), who required the class to read van der Waerden's classic text in the original German, as it had not yet been translated into English.

Looking back on Ambrose's teaching style, she says, "it seems to me now it was a rather ruthless attitude" to expect students to learn both German and algebra simultaneously. But, taskmaster though he was, "all the graduate students thought Ambrose was wonderful" (Cronin Scanlon: 16). Ambrose had held a postdoctoral fellowship at the Institute for

Advanced Study and had an infectious, almost missionary zeal for mathematical research.[10] In fact, Cronin Scanlon recalls that Ambrose "was more interested in the graduate students, I think, than in his colleagues" (Cronin Scanlon: 13).

Although Cronin Scanlon "really liked the work in the courses," she looks back on graduate school as "an awful lot of stress." The culture of graduate study at Michigan in the postwar years was one in which mathematical research reigned supreme. In this environment, "the chief anxiety which I had, and which I shared with everybody else, was . . . whether we really ought to *be* graduate students," she recalls. "Most of us, most of the time, thought we were *not* good enough!" (Cronin Scanlon: 12). There were several women among her classmates, and insofar as feelings of inadequacy were concerned, she says that "it never occurred to me that it was any different for the men than the women" (Cronin Scanlon: 12).

At the same time, the anxiety and pain of graduate school had an almost worshipful quality to it: the suffering was undergone for a higher good, for mathematics, to which Cronin Scanlon and her compatriots were as wholly devoted as acolytes. In the final analysis, it was the process of doing research—of proving something no one has ever proved before—rather than the degree itself, which appealed to her most of all and motivated her to continue.

During her last two years of graduate study at Michigan, Cronin Scanlon says, "it was decided that it would be a good thing for me to be doing some teaching" (Cronin Scanlon: 11). The idea of teaching, particularly at an elementary level, was not naturally very appealing to her. But the massive postwar surge of students had begun, and graduate student labor was desperately needed; moreover, she recognized that teaching experience might help her to get a job at some future time. What made these last two years worthwhile was that, at last, she began to do her own research. She worked in functional analysis with the German émigré Erich Röthe, who gave her "a very, very good problem" to work on (Cronin Scanlon: 13). Challenging, yet within her grasp, the problem she solved in her dissertation gave rise to many other questions and problems that would generate a fruitful line of research for years to come. Röthe had, in fact, set her on the road to becoming a bona fide research mathematician.

The Perils of Isolation: Jean Walton

During the early postwar years, with the demand for mathematics and mathematicians on the rise, graduate school was often a place where women as well as men could develop their mathematical skills and abilities. As the stories of Jane Cronin Scanlon (at Michigan), Mary Ellen Rudin (at Texas), and Maria Steinberg (at Cornell) serve to illustrate, the presence of helpful faculty mentors and the cameraderie of enthusiastic fellow classmates contributed greatly to their success. Still, even in this time of excitement and growth, it was possible for a woman to become profoundly isolated in her pursuit of the Ph.D.; such was the case for Jean Walton, who worked toward the Ph.D. at the University of Pennsylvania during the years 1945 to 1948.

Walton's thirteen-year path from bachelor's degree to doctorate is the longest among the interviewees who earned Ph.D.'s in the 1940s. On graduation from Swarthmore in 1935, she taught high school in New Jersey for three years. She then went on to work on a master's degree in mathematics at Brown, where "I was absolutely certain that I was *not* going on for a doctorate," she recalls. "I was really marking time until somehow or other this miracle happened and my social life turned around and I would get married" (Walton: 8). After earning the master's she was invited back to Swarthmore to work as an instructor and assistant dean. There she at last became deeply engaged in work that was meaningful and satisfying, and her life began to take on shape and direction. She acknowledged that it was fruitless to put her life on hold waiting for marriage; she decided that it was time to seriously pursue an academic career. Undecided as to whether to seek a teaching post or a job as dean, Walton realized that "either way . . . I've got to get a doctorate" (Walton: 12).

When, at age thirty, Jean Walton made the decision to work for a Ph.D. in mathematics, her former professors at Swarthmore recommended that she go to the University of Pennsylvania and work with the distinguished German number theorist Hans Rademacher.[11] Rademacher subscribed to a key tenet of the myth of the mathematical life course: that the development of a true mathematician must proceed without interruption. So at

Anne Lewis (Anderson) with Magnus Hestenes (left) and E.B. Shanks, 1962

Winifred Asprey, 1968

Lida Barrett, 1950

Grace Bates, 1996

Barbara Beechler, 1996

Jane Cronin Scanlon, 1996

Patricia Eberlein, 1997

Herta Freitag, 1995

Evelyn Granville, 1997

Susan Hahn, 1997

Anneli Lax, 1997

Edith Luchins in the 1990s

Margaret Martin, 1997

Cathleen Morawetz, 1983

Vivienne Morley, 1997

Mary Ellen Rudin, about 1990

Alice Schafer, 1973

Augusta Schurrer, 1996

Domina Eberle Spencer & Hypatia, 1998

Maria Steinberg, 1996

Rebekka Struik, 1997

Jean Walton, 1996

Tilla Weinstein (center) with Lipman and Mary Bers, 1991

his first meeting with Walton, he expressed incredulity at her desire to earn a Ph.D.:

I gave him a quick summary of [my background], and he said, "I do not understand! Either you are a mathematician or you are not a mathematician! Now you tell me you graduated with highest honors from Swarthmore in mathematics and then you *stopped?*" And I said, "Yes, that's what I'm telling you. I stopped." "And then you taught for three years, you went to graduate school and you got a master's degree in mathematics and then you *stopped again?*" And I said, "Yes. That's right. That's what I did." And I said to him, "Well, I guess by your terms I'm not a mathematician." And I said to myself, "And I think inside I'm probably not anyway," but I didn't say that to him. But I said, "I'm very eager to do this and I want to try." And so, he said, grudgingly, that he would let me try. And he took me on trial for that spring semester. (Walton: 13)

Rademacher, reluctant to accept a student who didn't fit his picture of what a mathematician should be, wasted no time in putting Jean Walton to the test. He asked her to read two papers in topology—one in French, the other a critique of the French paper, written in German. She was then to give a coherent presentation of both papers in his seminar, giving a careful assessment of the arguments in each. At first the task seemed overwhelming, requiring mastery of French, German, and a branch of mathematics that she had not previously studied. "I panicked inside," she says. "I thought, 'If my getting a Ph.D. depends on my being able to work all this out, I'm a dead duck. I can't do it'" (Walton: 14).

And yet, ultimately, she *did* meet Rademacher's challenge—and she did it entirely on her own. She didn't seek help from the students in the seminar; she didn't seek help from her colleagues at Swarthmore; she didn't seek help from Rademacher. Dutifully traveling from Swarthmore to Penn for the seminar, she returned home each time to plug away at the task he had set before her. The experience, Walton says, "did something for my confidence in myself that was very important, in that it said you can be totally bewildered about something and still work it out. You don't have to see through everything right away." What's more, Walton's independent effort impressed Rademacher, who accepted her as his doctoral student. At last, Walton recalls, "he became interested in me" (Walton: 15).

While passing Rademacher's test was a personal triumph for Walton, it set the tone for the rest of her stay at Penn. Her experience of doctoral study was essentially that of doing mathematics in seclusion, interrupted by periodic interactions with Rademacher. She developed no warm relations with other faculty, and she felt alienated from her fellow students: "I think it was partly the commuting, my sense of an age differential . . . and, very clearly, largely gender. . . . I had an *image* of what they were all like. I didn't really *know* them" (Walton: 22).

Certainly, Walton's tendency to stand apart from the other students, her intense self-reliance, her commuting from Swarthmore all played a role in the isolation she felt at Penn. But at the same time, in a graduate program that rapidly filling up with ambitious young men, women could easily be made to feel like outsiders who had to prove their mettle. A woman needed a strong sense of internal motivation and an irrepressible drive to do mathematics in order to emerge from graduate school with the strength to go on.

Walton continued to feel, as she had felt at Brown, that mathematics was simply "too remote from life" (Walton: 8). Although her dissertation was accepted for publication in the prestigious *Duke Mathematical Journal,* she remembers thinking that her research was of "very little interest" to other mathematicians (Walton: 22). Certainly Rademacher did little to introduce her to other number theorists with whom she could discuss mathematics and gain a sense of the significance of her work. The isolation she experienced in her graduate school years held her back from developing the sense of confidence and competence that comes from learning and doing mathematics with a sense of connection to a wider community.

The Ph.D.'s of the 1950s

The interviewees who earned Ph.D.'s in the fifties differ from those of the forties in two significant respects. First, the mean elapsed time from bachelor's to Ph.D. is dramatically longer in the fifties group: just under nine years. Second, the marriage rate in the fifties generation is much higher: seventeen of the nineteen women eventually married, and fifteen of them married before completing the Ph.D. While it would be an oversimplification to assert that the early marriage *caused* the slow progress to the Ph.D., it is clear that marriage and family life added complication—and sometimes conflict—to the graduate school ambitions of this later group of women.

The 1950s generation of women Ph.D.'s completed their graduate study during a time of cultural transition, in both the mathematical community and American society as a whole. At the end of World War II, American mathematics was poised on the brink of an era of explosive growth; the research community, in particular, enjoyed unprecedented prestige and excellent funding. At the same time, the wider culture placed a high value on the femininity of women, so that graduate study in any field—much less a manly field like mathematics—could be seen as dangerously defeminizing to the women who pursued it. In an odd way, despite all the complications it added to her life, marriage seemed to make the completion of the Ph.D. a more acceptable pursuit for a woman in the 1950s, since it signified her intention to take on the normal female roles of wife and mother.

In the sections that follow, I explore the unusual variety of graduate school experiences among the women of the fifties generation of Ph.D.'s. In particular, I examine how social and political forces, both inside and outside the mathematical community, affected those women who pursued the Ph.D. in the turbulent postwar era.

Exuberance and Ambivalence: Graduate School in the Postwar Era

The immediate postwar years were a tumultuous time for mathematics and science in American colleges and universities, a time of triumphant expansion and enthusiastic recruitment. Margaret Marchand describes the five years she spent in graduate school at Minnesota from 1945 to 1950 as having been "what I still consider the happiest years of my life" (Marchand: 6). She began her graduate study shortly after V-J Day, and the mood on campus was exuberant:

That first fall, the University of Minnesota had eighteen thousand students on campus. The second fall, they had twenty-seven thousand. And I remember Falwell Hall, the hall the math department was located in: they had stairwells marked for "up only" and "down only." In between classes, it was unreal. The place was jammed. . . . But everybody was working, and it was such an "up" time. The war was over—I think that was part of it—and all these people had an opportunity that they never thought they'd have. They'd survived the war. And I guess it was catching, because I remember feeling that same feeling, that "the sky's the limit!" (Marchand: 8–9)

She tended to view whatever was set before her—whether it was the courses she took, or the courses she taught, or the research suggestions of Robert Cameron, who had *invited* her to work on a Ph.D. with him— as a challenge to be happily met. If one word summarizes her attitude and demeanor during graduate school, that word is *gratitude.*

But as the years passed, the postwar exuberance gave way to a new normality. Triumph and prosperity meant that women could—and should—return to the home, bear and raise children, and in general, create a haven for their husbands and families; the public sphere was increasingly viewed as a place reserved for men. In her epoch-making book, Betty Friedan (1963) refers to the new paradigm for American women as "the feminine mystique" and asserts that it was decisively established as the cultural norm for American women by 1949. Not coincidentally, it was just about this time that the tidal wave of veterans who had gone to college on the G.I. Bill were ready to enter graduate and professional schools. A veritable army of bright and eager men stood ready to do battle in the fields of the intellect. What was it like for the minority of women who stayed on to complete a Ph.D. in mathematics during this time of radical social change?

Many of the women who began graduate study in mathematics very close to the end of World War II found themselves caught between paradigms as they worked to complete the Ph.D., as the graduate school stories of Augusta Schurrer and Barbara Beechler clearly illustrate. Augusta Schurrer began work toward the Ph.D. in mathematics at the University of Wisconsin in the fall of 1945. She had applied to graduate programs in physics, astronomy, and mathematics and had been accepted everywhere she applied. She attributes her success in the graduate school application process to the fact that "the war hadn't quite ended, so there was a dearth of men, and [the graduate schools] were particularly interested in getting students." She chose mathematics because she thought it offered more opportunities for women than physics or astronomy and because she wasn't strongly motivated to do laboratory or observatory research. Most of all, however, mathematics attracted her because she felt that there was still much for her to learn—that she could subsequently "share and teach" (Schurrer: 11).

For Schurrer, who had lived at home while at Hunter College, graduate school was the gateway to freedom and autonomy. "Somebody was going to *pay* me to continue studying," she says, "and they were going to pay me for teaching, which sounded pretty good to me. And I was going to be away from home, which at twenty-some was even better" (Schurrer: 12). Although she had done well at Hunter, she felt that problem solving was her strength and theorem proving her weakness. Her first semester at Wisconsin was a struggle, but she slowly mastered the art of proof by watching and studying the technique of the skillful teachers of the older generation, R.E. Langer and C.C. MacDuffee.

With her confidence somewhat shaken early on, it was a relief to know that the students coming in on the G.I. Bill were less well prepared than she. "I was in a safe position," she recalls. "People were filing in as graduate students who had definitely less academic ability than I and less training, basic training. So I was able to be of help to others" (Schurrer: 14). Like Jane Cronin Scanlon, Schurrer worried about whether she was "good enough" for graduate school, but unlike Cronin Scanlon, she seems to have kept this concern to herself.

On receiving her master's degree in 1947, she decided to stay on for the Ph.D. At about this time she was already beginning to sense a sea change in the department's attitude toward doctoral students. Younger faculty members were increasingly competitive when it came to research, and they were correspondingly more selective when it came to advising doctoral dissertations. In particular, she sensed that they were not interested in her because they "didn't want people who were going to get married off and leave" (Schurrer: 30). She briefly considered working on her Ph.D. with Langer or MacDuffee, but she felt that they were too close to retirement and already had too many students to be able to take her on as well. As one of the very few women working for the Ph.D. in mathematics at Wisconsin—she remembers just two others, only one of whom (Violet Larney) actually completed the degree there—she felt increasingly vulnerable.[12] It is possible that she felt a kind of survivor's guilt. "I saw what they did with women," Schurrer recalls. "I mean, women got cut at the master's very fast at that time. . . . They were just not encouraged to go on. They weren't men" (Schurrer: 29).

Casting about for an adviser, she finally selected Morris Marden of the University of Wisconsin at Milwaukee, who had come to Madison for a semester to give a course in complex function theory. Marden was "a *super* teacher" who had "a teacher's personality." He was working in complex analysis—a subject with which she had fallen in love while at Hunter College. Most important, perhaps, Marden was someone with whom she could feel at home:

Morrie was very patient with me and didn't have any other graduate students. He let me come up to Milwaukee and drink beer with him and talk math. . . . And he was from Brooklyn via Harvard, and, you know, we could relate to each other. (Schurrer: 17, 30)

Working with Marden took her out of the mainstream of the graduate program at Madison. While she enjoyed his company and his teaching style, she still struggled against what she perceived to be her own limitations:

I was very slow. I was having too good a time again to work. I couldn't find what I wanted to work on. I started a couple of topics, and they were not interesting, and there wasn't an awful lot available. And I didn't know how to be creative. I really hadn't had all that much math. I had basic algebra. I had basic, very basic analysis. And I realize now I didn't have many tools to work with. . . . I didn't have the tools that I needed. (Schurrer: 17)

It almost seems as if she needed to remove herself from the graduate school environment, where she felt out of place and inadequate to the task, to focus on the dissertation:

I went to visit a friend down in Ottawa, Illinois, over one break and finally resolved a theorem that had been bothering me. I was reading a book at night or something, thought about the theorem, and resolved the difficulty. And so then I was at the write-up stage. (Schurrer: 18)

While she was eventually able to complete the Ph.D., she essentially did so *in absentia,* taking a full-time job at Iowa State Teachers College in 1950 and receiving the degree in 1952.

Schurrer's difficulties at Wisconsin were twofold: first, her determination to study advanced mathematics so that she might "share and teach" was increasingly at odds with the competitive research environment she encountered there; second, her sense that women were unwelcome served to compound her sense of incompatibility with the school. The discomfort that Barbara Beechler felt at Iowa—where she was a student from 1945 through 1955, the last six of these years in graduate school—is far more difficult to sort out. Like Jane Cronin Scanlon at Michigan—and unlike Augusta Schurrer at Wisconsin—Beechler was very much caught up in the exhilarating research environment at Iowa. But over the course of her years there, she grew increasingly frustrated and disillusioned with research mathematics, for complicated reasons that seem intimately bound up with gender.

The University of Iowa produced only four women Ph.D.'s in mathematics prior to 1960, three of whom—Dorothy McCoy (1929), Winifred Asprey (1945), and Kathryn Powell Ellis (1955)—completed doctoral dissertations under the direction of E.W. Chittenden in analysis and point-set topology.[13] Chittenden had some ambivalence about women and the Ph.D.: when Winifred Asprey began graduate work at the University of Iowa during the war, he told her, "A Ph.D. isn't for a woman," but later he invited her to do her dissertation research with him. By all accounts he was a disorganized teacher but very loyal and committed to his students. His lecturing style, untidy as it was, provided his students "a beautiful example of how the research mind works" (Asprey: 20).

Barbara Beechler entered the University of Iowa as an undergraduate in the fall of 1945; Asprey had just been awarded her Ph.D. that August. As we have seen in chapter 4, Chittenden took a special interest in Beechler not long after her arrival on campus—his ambivalence about working with women apparently having given way to enthusiasm. She earned her bachelor's degree in 1949 and spent the next two years working on a master's thesis with him.

"Chittenden really did inspire me," Beechler recalls, although his style of teaching left a lot to be desired. "In all of the courses that I ever had with him, I never heard him complete a proof. If he gave the hypotheses of a theorem, he would not tell you the conclusion; if he gave the conclusion, you didn't know the hypotheses" (Beechler: 50, 17). But by filling in the gaps in Chittenden's presentation, Beechler learned how to formulate hypotheses and how to draw conclusions; she learned how proof *works:*

I thought that topology was grand, and I discovered axiomatics and wrote a paper for Chittenden in that topology course on axiom systems, topological axiom systems—whether they were consistent with one another or whether one implied the other and so on. And I thought that was great, you know, after all of that very formal stuff that I'd gotten in my logic class. (Beechler: 17)

A student could either be infuriated or energized by Chittenden's style of teaching; Beechler clearly fell into the latter camp. But her long-time companion, classmate, and fiancé, Bob Blair, did not; he found Chittenden's lacunae inexcusable. This was perhaps the first serious difference of opinion of their relationship. When Blair decided to break with Chittenden and go off in search of something and someone new, Beechler—accustomed to going along with Blair—eventually decided to follow.

On the face of it, this seemed to her like a prudent decision. Chittenden was the grand old man of a very small graduate program. "He was the highest-paid professor in the university," Beechler recalls, "and he drove a big Cadillac" (Beechler: 18). His specialty, point-set topology, while still an active research area, was not as fashionable as it once had been. The young Turks of the department at Iowa—Malcolm Smiley and H.T. Muhly—were doing exciting, timely research in modern algebra. Blair cast his fortunes with Smiley, and Beechler, following suit, chose Muhly as her dissertation director. In so doing, she left behind both a field and a mentor that had engaged her imagination and creativity, allowing herself to be persuaded that what was good for Blair was also good for her. She justified her decision not to work with Chittenden on the grounds that he was "too old, that he might not last." The irony of this rationalization became clear only later: "I did stand with Chittenden at Muhly's funeral [in 1966]! So much for my thoughts as a young person" (Beechler: 21).

At first the prospect of working with Muhly seemed like a new and exciting adventure. In the first algebra course she took with Muhly, she read van der Waerden's *Modern Algebra* in the original German, just as Cronin Scanlon had done with Ambrose in Michigan. She enjoyed the challenge of learning a new subject in another language, as she had done before with both Bergmann and Chittenden. But in time she came to realize that although she had immersed herself in algebra, she did not have the same intuitive feeling for it that she had had for Chittenden's topology:

I could prove the theorems. They were all nice, abstract theorems about characterization of ideals over Noetherian domains, and I worked in rings of polynomials in two variables and all these wonderful things. But they were just purely algebraic abstractions for me, and I could deal with it formally, but I didn't understand it. And I knew I didn't have anything to do with it. . . . I didn't feel engaged or involved. (Beechler: 31–32)

In short, Beechler never gained a sense of *possession*, either of the general subject matter or of the specific theorems she proved in her graduate research.

If Chittenden had been "too old" to direct her thesis, Muhly was, if anything, too *young*. "He didn't know how to pose a problem. He posed problems that were too hard. And he didn't know how to keep out of it himself," Beechler recalls; Muhly's interference made it that much more difficult for her to feel that the mathematics she was creating was truly her own (Beechler: 32). Making matters worse, she worked on her dissertation in isolation—physically, emotionally, mathematically—from practically everyone save Bob Blair, whom she finally married in the summer of 1951. Because the mathematics department at Iowa had a policy that prohibited male and female students from sharing office space, she and Blair sought an office together in the physics department. "We were together, essentially, twenty-four hours a day," recalls Beechler (Beechler: 31). Bob was progressing nicely on his own dissertation in commutative algebra, which became a central topic of their conversations.

By 1952, Blair had completed his Ph.D. and was awarded one of the first NSF postdoctoral fellowships to continue his research at the University of Chicago. Beechler had by this time completed her dissertation research but had not yet written it up; even so, Muhly felt that it was time for Beechler to "leave the nest" as well (Beechler: 22). Her search for a job landed her far away from both Iowa City and Chicago: it was, ironically, through her connection to Chittenden that she landed a temporary faculty position at Smith College in Massachusetts. The department at Smith was headed by a former student of Chittenden's, Neal McCoy, and it was Chittenden who recommended her for an opening there. During the two years Beechler spent at Smith, she and Blair gradually drifted apart, their marriage strained as much by their divergent mathematical fortunes as by the physical separation. By the time she finally received the Ph.D. in 1955, her marriage had ended, and she was quite thoroughly disillusioned by her doctoral research.

Barbara Beechler was a strong student, well thought of by all of her teachers at Iowa. Yet in a competitive environment defined and dominated by men, it is not surprising that she chose to follow the lead of the man who had been her childhood friend, closest companion, and husband-to-be. It is also not surprising that in this environment she chose an inexperienced adviser in a fashionable field over someone she perceived as being out of step with the changing times. But she questions the wisdom of the decision to this day.

Disruptions, Detours, and Delays

Among the interviewees of the fifties generation, by far the longest delays en route to the Ph.D. were those experienced by the European refugees, Herta Freitag and Susan Hahn, for whom the interval between first university degree and Ph.D. was nineteen and twenty-one years, respectively. Both women had prepared for careers in teaching in Europe and had their teaching credentials by the mid-1930s, but neither was able to secure a gymnasium-level teaching position in the thirties. Freitag worked as a mathematics tutor in Vienna until the *Anschluß* in 1938, whereupon she fled Austria for Great Britain and, ultimately, the United States. In England, she held jobs as a maid, waitress, and nanny and spent much of the war teaching in a school for displaced children.

On her arrival in the United States in 1944, Herta Freitag moved quickly to establish herself as a mathematics teacher, taking a position at Greer School, a residential school for disturbed children in upstate New York. It seemed natural to resume her own education at the same time: while visiting her brother in New York City in the summer of 1945, she enrolled as a student at Columbia Teachers College. During that summer, and for three summers thereafter, she worked toward a master's degree there, with Howard Fehr as her adviser.

The Ph.D. had not been her goal at the outset, but as the final summer drew to a close, Fehr urged her to go on for the doctorate. Freitag recalls that "when I told my friends they said, 'Listen, we have to fall on our knees to be accepted as doctoral candidates! You have *been* asked. By all means, do it!' And that really influenced me" (Freitag-2: 13).

Freitag's Ph.D., like the master's degree, was completed mainly at a distance and in the summers. Just as she began work toward the doctorate in 1948, she also resigned her position at Greer School and accepted a full-time position as instructor of mathematics at Hollins College in

Virginia. During the five years that she spent working toward the Ph.D., Freitag spent just one semester at Columbia to fulfill the residency requirement for the degree. Apart from that, she says, "I spent every free minute at home on my doctoral work, during my teaching time at Hollins" (Freitag-2: 18). During this same period, she married a former Greer School colleague, Arthur Freitag, who moved from Chicago to Virginia to be with her while she was in the thick of establishing a mathematics department at Hollins and completing the Ph.D. When at last, after eight years of graduate study, she finally defended her doctoral dissertation at the age of forty-four, it had come to seem like the fulfillment of a destiny long deferred.

By contrast, Susan Hahn's relationship to mathematics was more tenuous and intermittent. After completing a year of practice teaching in Budapest in 1937, she worked briefly at "voluntary" (unpaid) factory jobs; on her marriage in 1939, she became a full-time housewife. Her long hiatus from mathematics was due as much to the societal stigma against a working wife as it was due to the privations of the German occupation of Hungary during the war years and the Russian occupation afterward. She and her husband finally left Hungary in 1948, traveled through Germany, France, and England, and arrived at their final destination, New York City, in 1950. Arrival in the United States meant freedom in more ways than one: she quickly took a job working at the Hanover Bank. "I was glad to get a job," she recalls, "but I wanted to be a mathematician. And I thought, for being a mathematician in America, I needed an American degree" (Hahn: 14). Save for brief stints as a teacher in France and England in 1949, Susan Hahn had been disconnected from academic life for fourteen years when she enrolled in graduate coursework at NYU in the spring of 1951.

The delays of Herta Freitag and Susan Hahn seem to have an heroic quality, shaped as they were by monumental events on the world stage. But the educational twists and turns of the other women who experienced long delays en route to the Ph.D. are no less heroic. The politics of the McCarthy era created daunting obstacles to the educational progress of Rebekka Struik. She recalls, "I think getting my Ph.D. was almost half an accident" (Struik: 5). On completing her bachelor's degree at

Swarthmore in 1949, she says, "I didn't want to continue. I just had had too much of books and study and wanted to do something else for a year or two" (Struik: 8). Seeking a change of pace, she sought employment as a computer programmer. After a first, unsuccessful job in programming in Philadelphia, she landed a more promising position at the University of Illinois. The programming she did there required a level of mathematical sophistication slightly beyond that of her undergraduate degree, so she began the practice of taking one graduate course in mathematics each semester, thus easing her way back into academic life.

But while working and studying at Illinois, she became active in leftist political causes; in particular, she worked with the Young Progressives of America to circulate the Stockholm Peace Appeal. Her political activities led to her dismissal from the programming position at Illinois.[14] By this time it was clear to her that her employment experiences weren't going very well; so she became a full-time master's degree student at Illinois and decided, on the advice of her professors, to go on for a Ph.D.

In 1951 she was accepted into the doctoral program in mathematics at Northwestern University with financial aid. But on her arrival on the campus the offer was withdrawn on political grounds: her dismissal from the job at Illinois was said to have been based on her involvement in "Communist activities" (Struik: 10). This turn of events left her scrambling for a way to continue living in Chicago and going to graduate school. Fortunately, she was promptly admitted and offered an assistantship at the University of Chicago—her *third* graduate school.

But it was not to be at Chicago that she earned the Ph.D. In July of 1952 she met Hans Freistadt, a physicist on the faculty of the Newark College of Engineering, while he was visiting mutual friends in Chicago; they had common interests in science and in progressive politics, and they wed in December of that year.[15] And so she moved to Newark, with the intention of resuming work toward the Ph.D.

She considered both NYU and Columbia, ultimately choosing NYU because their late-afternoon and evening classes left her time for part-time work. At NYU—her *fourth* graduate school—Rebekka Struik was at last able to complete her doctoral studies. After the many disruptions of her previous graduate school experiences, the dislocations incurred by night school, commuting, and working part-time may have seemed minor by comparison.

For a few women of the fifties generation, the time from bachelor's degree to Ph.D. was prolonged by a natural process of experimentation— of trying on a variety of personal and professional roles until they found the proper fit. The experience of Patricia Eberlein, who earned her doctorate eleven years afer completing the bachelor's degree, is noteworthy for its extraordinary geographical, vocational, and personal variety. As an undergraduate at Chicago in the 1940s, her interest in creative writing and her desire to "learn truth" tempted her to consider a major in English or philosophy. But ultimately, she majored in mathematics because "it was the one thing I was able to do" (Eberlein: 3). By the time she completed her bachelor's degree in 1944—having already experienced a brief marriage that ended in divorce—she spent a summer in Mexico and then moved to New York City to seek her fortunes as a creative writer:

I had a whole series of jobs: as a photographer's assistant, at one point; writing a children's encyclopedia, in another; working for Twentieth Century Fox, writing synopses of novels so that someone could pick up the gist of it and decide whether or not it would make an appropriate movie script. And I got stuck on Henry James and swore I'd *never* do another one of those! (Eberlein: 5)

In the spring after the end of the war, she joined a group of New York friends on an automobile trip to the West Coast. On the trip west, she met and married a young man and settled down to a ranching life with him in South Dakota. They had two children; eventually, she says, they "went broke" (Eberlein: 5). At the urging of her parents, who were then living in Michigan, they moved the family to East Lansing in 1949. Her husband, who had never been to college at all, enrolled in a program in animal husbandry at Michigan State, where his tuition was paid by the G.I. Bill. Eberlein, on the other hand, had no intention of returning to school; but when she looked for a job, she discovered that she could earn more money as a graduate student paper grader than she could working full-time anywhere else. So she became a graduate student in mathematics and in time qualified for a graduate assistantship. "And then," she says, "I was just hook, line, and sinker into mathematics" (Eberlein: 6).

Graduate School as Obstacle Course: The Case of Lida Barrett

Patricia Eberlein came to graduate study in mathematics after a lengthy journey of self-discovery that included marriage, motherhood, and a variety of livelihoods and lifestyles. Once she decided to work for the Ph.D., however, issues of marriage and family did not seem to impede her progress toward the degree. But the fifties generation does include a number of women for whom the needs of spouses and children, and discriminatory social attitudes toward wives and mothers, significantly disrupted their efforts to complete the doctorate.

When Lida Barrett received her bachelor's degree from Rice Institute in 1946, she was just eighteen years old. In little more than eight years' time, she would hold a succession of jobs, begin work toward a doctorate at one graduate school, marry, try to start a family, and ultimately complete her Ph.D. at yet another graduate school nearly two thousand miles away from the first. Her journey from bachelor's to doctorate was delayed by institutional prejudice against married women but at the same time expedited through the efforts of a supportive husband and—ironically—by the unexpected difficulties they encountered in their attempts to conceive a child.

At the time of her graduation from Rice, Barrett recalls, the war had just ended, jobs in Houston were plentiful, and "the last thing I wanted to do was go on to school" (Barrett: 7). Instead, she went to work for Schlumberger, the oil well service company, and continued to live at home with her family. It rapidly became apparent, however, that her future at Schlumberger would be limited without a graduate degree. Not quite ready to return to student life but ready to move out of her childhood home, she left Schlumberger in the late summer of 1947 and accepted a last-minute offer of a teaching position at Texas State College for Women (now Texas Women's University) in Denton.[16]

At Denton, Lida Barrett came under the influence of Harlan Miller (Ph.D., Texas, 1941), the first woman to complete a Ph.D. under R.L. Moore at Texas.[17] As head of the department, Miller conspired to hasten Barrett's entry into graduate school in every way possible. First, she lured Barrett to the teaching position at Denton with the offer of a private graduate course in elementary set theory. Then, at the end of the academic year 1947–48, she persuaded Barrett to spend the summer in Austin taking a summer course in point-set topology at the University of Texas. On the strength of her summer school performance, Texas offered Barrett a graduate assistantship for the academic year 1948–49. At first, she balked at the offer; she had planned to return to Denton for a second year of teaching. But Miller foreclosed that option, saying, "You have absolutely no point in coming back here. You need to go to graduate school" (Barrett: 9). And so, in the fall of 1948 at the age of twenty-one, Barrett began graduate work in mathematics at Texas.

In the lively graduate student milieu at Texas—where she and Mary Ellen Rudin were the only female students—Barrett quickly made herself at home. In her first year, she wrote a master's thesis with H.S. Wall; she decided to work on a Ph.D. with R.L. Moore; and she met her future husband, John Barrett, who was one of her fellow graduate students. Although she was working toward a doctorate in pure mathematics, she sustained a strong interest in applied mathematics by continuing to work for Schlumberger in the summertime. In fact, during her second year at Texas (1949–1950), she declined an assistantship in the mathematics department, working instead for the Defense Research Lab on campus. "I could have gotten a master's in physics for the work I did," Barrett recalls, "but . . . I didn't want to take the time to write it up" (Barrett: 12).

Lida Barrett was next in line for a fellowship from the mathematics department at Texas when she and John announced their engagement in the spring of 1950. This seemingly innocuous announcement gave rise to a frustrating sequence of events: "I got moved to an alternate's position because if I was getting married, they weren't going to waste an assistantship on me." Her response to this blatantly unfair decision, even fifty years after the fact, is curiously ambivalent. On the one hand, she says she "wasn't too unhappy about" losing the fellowship at the time: "My plan was that, as soon as John had finished his doctorate, we would go ahead and have a big family and *then* I'd go back and finish my doctorate." Losing the fellowship must have upset her nevertheless, for she is quick to point out that "The two men who got the [fellowships] both got doctorates, but they got them *after* I got mine" (Barrett: 11).

Although the fellowship opportunity was denied her, she could console herself in the pleasure (and remuneration) she continued to draw from her work at the Defense Research Lab. But when she married John Barrett later that same year (1950), she lost that job, too. "My husband was a full-time instructor," Lida Barrett recalls. Although her position was only half-time and "had absolutely nothing to do with the math department," she was dismissed under the antinepotism rules then in force at Texas (Barrett: 12). As so often happened, it was the woman—not the man— who lost her job.

Having lost both a fellowship and a job because of her marital status, Barrett's frustration became more difficult to contain. The decision to get married had been predicated on the assumption that both she and her husband would be earning an income. With the loss of the job at the lab, Barrett turned to tutoring as one of the few remaining money-making options. Moreover, Barrett says, "if I had stayed at the Defense Research Lab, I probably would have written up the research and had the master's" in physics. But she continued to try to make the best of a bad situation. Losing her half-time job meant she would have more time to concentrate on mathematics. "The whole business of not getting the fellowship," she says, in hindsight, "and getting booted out [of the lab] was just part of the culture of the time" (Barrett: 12).

The academic year 1950–51 was her final year in Texas. John Barrett received his Ph.D. at the end of that year and, in a difficult job market, accepted a position at the University of Delaware. While R.L. Moore hadn't defended her right to the fellowship in Texas, he did not hesitate to refer Lida Barrett as a student to J.R. Kline at the University of Pennsylvania.[18] During the fall of 1951, however, her main concern was the effort to conceive a child. "I just took two courses and wouldn't take an assistantship," Barrett recalls. But as time went by and she wasn't getting pregnant, she became more seriously committed to completing the Ph.D. In the fall of 1952, she accepted the assistantship and became a full-time student at Penn. Boarding the Baltimore-to-Wilmington commuter train in Newark, it was "two stops to Wilmington, run down stairs and up the other stairs, and twenty-two stops to Philadelphia and then take the trolley [to Penn]. . . . I left home at 7:10 and got to the campus around 8:30." On the return trip in the evening, she took "a B&O express train, that made one stop in Wilmington and paused in Newark" (Barrett: 13).

Barrett was considered an odd duck among the graduate students at Penn. As much as possible, she tailored her classes and exams to the train schedule; as a consequence, she rarely consulted her classmates or professors with problems or questions. Her husband John routinely came up to Penn for a Wednesday night seminar; he would use the library, and they would eat out with friends on the faculty whom they'd known at Texas. Her comings and goings, and her social life with the faculty, made her mysterious and "sort of different" (Barrett: 15). Because she lived inconveniently far from the Penn campus, much of her work toward the doctorate was completed independently. But she had considerable personal resources to fall back on in times of difficulty: she relied on the deductive skills she had developed as a student of Moore at Texas and—in a pinch—on her husband's expertise. These resources prevented her from experiencing the damaging sort of isolation that had marred Jean Walton's doctoral study at Penn just a few years before.

Lida Barrett finally earned the Ph.D. in 1954, shortly after her twenty-seventh birthday. That she did so despite so many obstacles—repeated instances of sexism, geographical and personal dislocation, the frustration of infertility—is a testament to her determination and persistence.

Marriage and Mathematics in the Fifties

For all of the detours she was forced to take because of her marriage, Lida Barrett's route to the Ph.D. was remarkably swift. Certainly, one of the key factors that enabled her to finish as quickly as she did was the remarkable partnership she enjoyed with her husband, which was extraordinary for its time and place. "He kept house," she recalls. "He picked me up every night at the station and already had supper ready. . . . He was *very* insistent I finish my degree. He said he thought I was as good a graduate student as he was; it was silly not to get the degree. I don't know where I found such a liberated man in 1950!" (Barrett: 16)

Indeed, many of the married women who earned Ph.D.'s in mathematics in the 1950s did so at the urging, and with the material and emotional support, of their husbands, as the experiences of Barrett, Joyce Williams, and Joan Rosenblatt clearly illustrate. In December of 1949, having just earned bachelor's degrees from the University of Minnesota, Joyce Williams and her husband married and immediately applied to graduate schools—he in physics, she in mathematics. "He got an offer of an assistantship at Illinois," Williams recalls. "That's why I went there" (Williams: 5). She, on the other hand, *was not* offered an assistantship at Illinois for her first year. But the first year (1950–1951) at Illinois was not a financial strain: "We had the G.I. Bill and his half-time pay. That took care of things" (Williams: 6).

In an ironic twist, her husband's assistantship was not renewed at the end of his first year in the graduate program in physics at Illinois, while the mathematics department offered Joyce Williams a fellowship that lasted for three full years. In a move uncharacteristic of young couples of the early fifties, her husband took a job in Buffalo while she remained at Illinois to finish her degree. "We traveled back and forth quite a bit," she recalls. "It wasn't an ideal situation. I mean, I offered to go with him, [but] he thought I should stay and finish my degree. So I did" (Williams: 11). For nearly three years, they had a commuter marriage more characteristic of the 1990s than the 1950s, and they waited to start a family until she received the Ph.D. in 1954.

Joan Rosenblatt began her graduate work in mathematical statistics at the University of North Carolina in 1948, having spent the two years following her graduation from college working for the National Institute of Public Affairs and the Bureau of the Budget in Washington, D.C. During her second year in residence in Chapel Hill, she decided to marry David Rosenblatt, whom she had met at the Budget Bureau. In fairly short order, she selected a dissertation adviser, Wassily Hoeffding; scheduled her preliminary oral examination for the Ph.D.; passed it; left the campus; married; and joined her husband in Pittsburgh, where he was then living. Living first in Pittsburgh and then back in D.C., Joan Rosenblatt tried to maintain a sense of connection to her graduate work by enrolling in mathematics courses but found herself losing focus. Ultimately, it was her husband—who had neither a master's degree nor a doctorate of his own—who convinced her to return to North Carolina and work on the Ph.D. in earnest:

He was the one who pushed me into getting back into serious pursuit of my degree. He pushed me, and I commuted to Chapel Hill for about a year and a half. I came back to Washington every two weeks. And once or twice, my husband came down. I was just keeping my nose to the grindstone. . . . He pushed me very hard: "Hey, you've gone this far! For heaven's sake, finish it!" I was glad to be finishing it up. (Rosenblatt: 10–12)

Marriage could be both help and hindrance to graduate study, and usually a combination of the two. Marriage—particularly marriage to a fellow mathematician or academic—could help to alleviate the loneliness and isolation that women could feel in graduate school. At the same time, a husband's professional ambitions often took top priority, and responsibility for home and family often fell disproportionately to the wife. Under such circumstances, it was often difficult for a woman to embark on graduate study without numerous delays, detours, and other changes in plan.

Edith Luchins's progression from bachelor's degree to Ph.D., which took a total of fifteen years, provides a case in point. Luchins, like Lida Barrett, married in her early twenties. Luchins encountered comparatively little of the structural discrimination that had impeded Barrett's progress, but she succeeded, where Barrett had failed, in starting a large family early in her marriage. The fulfillment of her dream of a Ph.D. in mathematics required a spectacular act of balancing the not always compatible needs of household, children, and a spouse's career, while at the same time maintaining an unwavering vision of her ultimate personal goals.

Luchins graduated from Brooklyn College in the spring of 1942 and married the following October. Inspired by a sense of patriotic duty, she spent the academic year 1942–43 engaged in war work, first for the British Purchasing Commission and later for Sperry Gyroscope. At the same time, however, determined to go on in mathematics, she applied to graduate programs at Brown, Johns Hopkins, and NYU. When her husband entered the army, they decided together that it would be best for her to remain in New York; so in 1943, she began graduate study at NYU. When she received the master's degree there a year later, she says, "It was just taken for granted that I would go on for the Ph.D." At the very same time, she was offered an instructorship at her alma mater, Brooklyn College, which she felt was simply too good to pass up. Although Richard Courant, who had taken her on as his doctoral student, regarded her teaching job as an irritating diversion, Luchins was able to balance the demands of teaching and graduate study very well: "I did not neglect my courses at NYU; I worked diligently on them and earned an A in every class" (Luchins: 16).

When her husband returned from overseas and they decided to start a family, however, the balancing act became significantly more difficult. When she informed Courant that she was pregnant with her first child in the spring of 1946, "he banged the table, and he exclaimed, 'But that's an interruption!'" (Luchins: 17). For a time, she persevered in both teaching and research, until to do so no longer seemed appropriate. "The social mores raised questions and eyebrows," Luchins recalls. "You were in your fifth month, and you were traveling on the subways to go to work? It didn't seem like the right thing to do." In hindsight, however, she says, "I should not have interrupted my education at NYU," and feels confident that "NYU would not have objected to my going there while I was pregnant" (Luchins: 18). But in view of Tilla Weinstein's experience as a pregnant undergraduate student at NYU nearly ten years later, perhaps Edith Luchins's confidence in NYU was misplaced.

While her husband worked two jobs—chairing the psychology department at Yeshiva College and working as clinical psychologist for the Veterans' Administration—Edith Luchins stayed at home with her newborn son, David. A second son, Daniel, was born less than two years later. While she enjoyed being home with her children, she clearly missed teaching and graduate study:

They seemed much easier than what I was doing! . . . I got very ambitious, and the fall after Daniel was born [in 1948], I started teaching an evening math course at Brooklyn College. I also signed up for a course at NYU. But it turned out to be too much for me to do all that and care for a toddler and a baby and an apartment, too. I developed a severe case of eczema on my hands, and I reluctantly dropped the course at NYU. I thought that fewer people would be hurt that way than if I had decided to stop teaching the class at Brooklyn College. (Luchins: 18)

This departure from NYU after the birth of her second son proved to be her last. In 1949, she moved with her husband to Montreal, where he had accepted a job at McGill. In Montreal, however, there would be no opportunity for Edith Luchins to return to graduate study in mathematics because McGill did not have a Ph.D. program. "I had essentially finished my coursework at NYU," Luchins recalls. "I'd taken the language exams, and I was about to take the comprehensive exam, but I didn't. I always knew I wanted to go back and finish. I had a great urge to finish" (Luchins: 21). While the realization was a disappointment, she made the most of the opportunity to enjoy her family and its connections to the vibrant life of the Jewish community of Montreal.

During these years, however, she had by no means abandoned the life of the mind. With her dream of completing the Ph.D. in mathematics on hold, her intellectual fulfillment came through collaboration with her husband. Their joint work combined his expertise in educational psychology with her interest in the teaching of mathematics and helped her to keep alive the dream of one day finishing the doctorate.

Oddly enough, political events conspired to return the Luchins family to the United States and enabled Edith to return at last to graduate school:

Congress passed the McCarran-Walter Act in 1952, which said that if you had not been born in the United States—if you had "derived citizenship," which I had—you would lose your American citizenship if you stayed out of the country for more than five years, unless you came back for eighteen months to the United States.[19] However, to come back for eighteen months, one needed to have a job! And so we reluctantly left Canada. (Luchins: 21)

In 1954 her husband accepted a position at the University of Oregon, and it was there—six years after her last attempt to take a graduate course at NYU—that Edith Luchins resumed her work toward the Ph.D. By the time she finally earned the degree in 1957, she was not quite thirty-six years old. She had already had four children and would give birth to a fifth the following year. She had packed a great deal of life experience into the fifteen years that had elapsed since her college graduation.

Just as Edith Luchins was completing her Ph.D. at the University of Oregon, Tilla Weinstein and her husband were creating a remarkably equitable marriage as they both worked on their doctorates at NYU. It certainly helped that they were working on their doctorates on very similar timetables, and living in close proximity to both sets of parents during the early years of their marriage helped to relieve the burden of child care. "He *appreciated* what I was doing, he *approved of* what I was doing, without any self-conscious sense of there being something special about it," says Weinstein of her husband. "*He* wanted to go on [to the

Ph.D.]; he saw no reason why *I* shouldn't go on. We just shared the care of the child" (Weinstein: 11).

At the same time, she was the recipient of a most unusual fellowship. Where just a few years before, Lida Barrett had been denied a fellowship at Texas because she was married, Tilla Weinstein received a National Science Foundation fellowship in 1955 that not only paid her own tuition and expenses but provided financial support for her husband and infant child:

> I was impressed by the fact that they were as helpful to me as they were in a time when I don't think other institutions took the same chances on women. . . . I forget the language. I think if you were married there was an extra allowance, simply because you were married. And there were dependency allowances for children. . . . And in other places, it was so clearly understood that the extra money was for a wife, that only husbands needed to support children, that, for instance, at the Institute for Advanced Study at the time, a visiting member would be given a different sum if they were married, if they were male—not if they were female. And this didn't change until the seventies. (Weinstein: 11–12)

In the final analysis, it was an extraordinary confluence of forces—family and spousal support; the forward-looking practices of the NSF; the advocacy of her adviser, Lipman Bers; her ability and sheer determination to succeed—that enabled Tilla Weinstein to complete the Ph.D. in a timely fashion.

For those women who combined marriage and motherhood with doctoral work in mathematics in the 1950s, the years of graduate study were far from the uninterrupted, single-minded, monastic apprenticeship spelled out in the myth of the mathematical life course. These women had to create their own opportunities and environments for concentration and creative effort—to create a niche in the complicated ecosystems of their lives for the pursuit of the Ph.D.

The Family Atmosphere of NYU

Many of the interviewees who earned Ph.D.'s in the fifties were so preoccupied with the logistical challenge of balancing their personal lives and professional goals that they had little time to be involved in the social aspects of graduate study. But there were a few noteworthy mathematics departments where women—married as well as unmarried—were made to feel part of a community of shared intellectual endeavor, and these institutions had a lasting impact on the women who earned their Ph.D.'s there.

New York University was the leading producer of women Ph.D.'s in mathematics during the 1950s. The success of women at NYU was due in no small part to its environment of intellectual excitement, which nevertheless accommodated the complexities and diverse commitments of women's lives. A model urban graduate program, nearly all the graduate courses and seminars at NYU were given in the late afternoon and evening, accommodating those women and men who cared for children or worked at paying jobs by day. Tilla Weinstein recalls that many of the graduate students, male and female, were married; nearly all of them commuted to the institute from one of the five boroughs or the near suburbs of New York or New Jersey. Susan Hahn recalls late night travels on the subway between Manhattan and Queens: "To go home at eleven o'clock on the subway in those years was no problem, absolutely no problem" (Hahn: 14).

The environment at NYU was a nearly seamless blending of work and study, research and instruction, the personal and the professional. Cathleen Morawetz and Anneli Lax first came to NYU as employees. Morawetz was hired for her expertise in electrical engineering, after completing a master's degree in applied mathematics at MIT; Lax was hired by the aeronautics department to compile bibliographies and carry out calculations during the war. Both of them subsequently became involved in the preparation of the celebrated shock-wave manual of Courant and Friedrichs. Employees of NYU routinely took courses and seminars there, and it was by this route that both Morawetz and Lax became students and worked toward the Ph.D.

There was a mood of excitement at NYU in 1946–47, the first academic year untouched by war. Friedrichs, in particular, was especially pleased to be teaching a topology class, his first pure mathematics course in several years. Cathleen Morawetz was caught up in the excitement:

I thought I'd take a course, too. Now, I didn't intend to go on in graduate school; I felt that that wasn't such a good idea because my husband had gotten a master's degree but he wasn't going to go further. But I got swept up by the—well, it was a very stimulating atmosphere that I had not experienced at either MIT or

Toronto. And *that's* when I became interested in mathematics. Of course, my own personal life had settled down more, so that also made a difference. But it was a lot of fun. There weren't so many full-time graduate students: there were people like me, who were working on projects right within the group. And there was a really very wonderful, enthusiastic environment—a lot of mixing between the faculty and the students. And so it was a great time. (Morawetz: 10)

It was also not uncommon for graduate students who had jobs elsewhere to ultimately become Courant's employees. In the early fifties, Susan Hahn, working by day at a Manhattan bank, enrolled in evening mathematics classes at NYU. In time, however, Hahn sought a job at the institute—because institute employees had better access to the library and better library privileges than mere students did!

Anneli Lax married a classmate from her complex analysis class, Peter Lax, who completed his Ph.D. several years before she did and joined the faculty. Marriages among faculty, staff, and students were fairly common. Tilla Weinstein recalls that, in contrast to many other institutions of the time, "there was a kind of nepotism that usually doesn't work but that worked beautifully at NYU" (Weinstein: 16). If, at times, NYU's mathematics program felt like one big family, it is perhaps because, in significant respects, that is exactly what it was.

In Tilla Weinstein's view, "one of the nice things" about NYU was that "there were enough women there so that I had women friends in graduate school," something she had not enjoyed as an undergraduate there (Weinstein: 16). Under Courant's leadership, the institute at NYU was, indeed, extraordinarily supportive of women. Cathleen Morawetz bore two children while in graduate school; during the first pregnancy, Courant gave her what was "essentially . . . a leave of absence" from her position at the institute. Morawetz recalls that when she returned a few months later, "he [Courant] gave me a raise so I could pay for the help that I needed at home." These provisions and arrangements "were all informal, wonderful things that you don't get nowadays"—provided through the beneficence of Courant (Morawetz: 10).

While the family atmosphere at NYU clearly fostered the participation of women in an intellectual effort, it had much of the hierarchical structure of the stereotypical family of the fifties with a strong patriarch at its head. Within the warmth of the NYU family, there was something of a gendered division of labor. It was not uncommon for women graduate students to be assigned tasks that were editorial or clerical in nature. For example, as Courant's assistant, Susan Hahn translated textbooks and monographs from English to German. Courant gave similar assignments to the wives of professors at the institute, and Hahn reports that he rarely gave proper credit for such work.[20] Over time, however, women usually graduated to more explicitly mathematical and scientific work. Such was the case for Cathleen Morawetz and Anneli Lax, both of whom did scientific research for Courant and Friedrichs with funding from various military and government grants.

Both men and women who worked in applied mathematics at Courant Institute were often encouraged to stay at NYU after completing the doctorate. The men who stayed on after completing the Ph.D. often rose to tenured positions of power and authority of their own. They assumed the same sort of role as the sons in a family who join the family business with a view to running it themselves one day. Female students, by contrast, were typically offered research associateships rather than tenure-track positions in the institute once their doctoral work was completed.

In important respects the women who did their doctoral research in pure mathematics were given a better preparation for leaving the nest and becoming a part of the mathematical community beyond New York City. Nearly all of these women worked with Lipman Bers or Wilhelm Magnus. Rebekka Struik recalls, "I liked going to Magnus every week with whatever I had done toward my thesis." She remembers "being very disappointed" when, after just a few months of work, Magnus told her that her dissertation was essentially complete. She had wanted to continue working with him, but "he didn't have the time. I think he simply had one student after another. As soon as he saw somebody had a respectable thesis, he tried to get them disposed of so he could help the others. That's what I think may have been going on" (Struik: 12–13).

Bers, by contrast, did a great deal for his students—female and male—to ease their transition into a mathematical life beyond NYU. He advised Tilla Weinstein, for example, to work on two separate problems for her dissertation, telling her, "If you come out of graduate school with two theorems rather than one, that will really set you apart" from other new

Ph.D.'s (Weinstein: 20). In the family structure that was Courant's institute, Bers may well have been the perceptive uncle who recognized the importance of preparing young people to strike out on their own.

Moving On

The women who earned Ph.D.'s in mathematics in the forties and fifties found themselves in a paradoxical position once graduate school was behind them. They were highly educated women prepared for research, teaching, and other forms of professional employment who found themselves in a society that expected them to find satisfaction as wives and mothers. Yet they were, by and large, women who sought to live fulfilling lives in both the public and the private spheres. The Ph.D. was a critical milestone in their lives, achieved through years, sometimes even decades of perseverance. While in some respects it seemed that the most critical hurdle was behind them, another, greater challenge lay just ahead: how to develop into a professional mathematician, skillfully interweaving a career and a life.

6

Interweaving a Career and a Life

Starting Out

While earning a Ph.D. seems a monumental achievement in itself, it is not so much an end as a beginning. According to the myth of the mathematical life course, the first few years after graduate school are absolutely critical as the new Ph.D. establishes the research program that will ensure his (or her) place in the mathematical community for years to come. To get a mathematical career off to the best possible start, the claims of marriage, family, and personal life must necessarily be subordinated to those of mathematics.

But for women, such subordination is neither possible nor, in many cases, desirable. Women conceive and bear children, and, in American society, they have primary responsibility for child care. Unmarried or childless women are often called on to care for aging parents. Thus it is not surprising that, for the women in this study, the major task during the early postdoctoral years and beyond was to find a way to balance the often contradictory claims of personal and professional life.

Moreover, not all of the women who earned Ph.D.'s in mathematics during the 1940s and 1950s aspired to the sort of single-minded research career outlined in the myth. While many of them sincerely loved research and aspired to research careers, some also wanted to be teachers or scholars in a broader sense than that spelled out in the prevailing model of the postwar era, and some aspired to careers in government, industry, or academic administration. Furthermore, several of the interviewees already had husbands and children by the time they earned the doctorate, and for them, the challenge was to find a way to make mathematics an integral part of their lives while at the same time fulfilling their ongoing responsibilities at home.

Nine of the women interviewed never married, and a tenth (Beechler) remained single after a brief early marriage. None of these women identifies herself in the interview as gay or lesbian, but many of them formed close emotional connections to other women, and all of them drew support from networks of male and female friends and colleagues. The problem of balancing the demands of career against the needs and requirements of personal life was no less challenging for these women. While it is true that they frequently felt greater freedom to travel and to exploit career opportunities, their professional lives were complicated by the fact that their colleagues, and society as a whole, tended to view them as carefree (even irresponsible) individuals who had much more time to take on more work and at the same time required less pay.

What were the strategies employed by the women of this generation to establish and maintain a sense of professional identity in the years following receipt of the Ph.D.? And how did these women manage to maintain and to nurture their personal lives in the process? In this chapter, I look at the opportunities and the obstacles, the progress as well as the setbacks, experienced by the thirty-six interviewees as they struggled to establish a *modus vivendi* for their personal and professional lives.

The Old-Boy/Old-Girl Network and the Women's Colleges

Many, though by no means all, of the women who earned their mathematics Ph.D.'s in the early 1940s proceeded to their first academic position in mathematics by means of what I have referred to in chapter 2 as the old-boy/old-girl placement system. In his history of the mathematics department at Chicago, Saunders MacLane (1989) asserts that, prior to the end of World War II, this was the mechanism whereby nearly *all* jobs in mathematics were secured. How did women actually obtain teaching jobs through this system? When seeking to fill a faculty vacancy, department heads at the women's colleges often turned to the institutions from which they had earned their Ph.D.'s, seeking women with fresh Ph.D.'s to join their staffs. On occasion, it would happen that a woman who was resigning from a faculty position at one of the women's colleges—often, though not always, because of her impending marriage—would take it on herself to assist in finding a replacement. She, too, would turn to those graduate departments where she had connections and contacts and inquire about the availability of newly minted female Ph.D.'s (Rossiter 1982: chap. 1, 1995: chap. 10).

The old-boy/old-girl network was quite limiting in the opportunities it provided for women Ph.D.'s. On the demand side of the system, women's colleges generally sought to fill faculty positions with women scholars; research universities generally sought to fill their positions with men. But on the supply side, women were rarely *recommended* for positions at coeducational institutions. As MacLane matter-of-factly asserts, "Chicago did not normally send its women Ph.D.'s to universities anxious to acquire research hotshots" (1989: 145). The prejudices on both the supply and demand sides of the system tended to reserve opportunities at coeducational liberal arts colleges, comprehensive universities, and research institutions for male Ph.D.'s.

This bias seems especially remarkable in view of the fact that temporary research positions, including National Research Council postdoctoral fellowships and visiting memberships at the Institute for Advanced Study, *were* awarded to women in the 1930s. It is as if the community deemed it acceptable for a woman to pursue research at the very beginning of her career but did not expect her to continue for more than a year or two beyond the Ph.D.[1]

Mary Dean Clement, Alice Schafer, Anne Lewis Anderson, and Winifred Asprey launched their teaching careers through the old-boy/old-girl system.[2] In each case, the hiring institution was a women's college, and a woman mathematician from the pre-1940 generation of Ph.D.'s played a key role in the hiring decision. Only rarely did the hiring process involve an actual interview or a formal application.

The recruitment of Mary Dean Clement to the faculty at Wells College, a small women's college in western New York State, gives a sense of the casual informality of the system. During the first half of 1941, Dorothy Manning (Ph.D., Stanford, 1937) paid a visit to the University of Chicago. She was about to marry Malcolm Smiley (Ph.D., Chicago, 1937) and was seeking someone to replace her as instructor of mathematics at

Wells.[3] Clement was the only woman completing her Ph.D. at Chicago that year, and Manning interviewed her on the spot. On the strength of Manning's recommendation, Clement recalls, "I was offered the job sight-unseen" (Clement: 22).

For Alice Schafer, who earned her Ph.D. the following year, the Chicago placement system proceeded more formally. In 1942, three women's colleges—Smith, Randolph-Macon, and Connecticut—contacted Chicago seeking new Ph.D.'s for their faculty. Schafer, the only woman finishing the Ph.D. that year, was a candidate for all three positions, although that fact was kept entirely secret from her throughout the process. The Chicago placement office took the liberty of declining the position at Smith *for* her, without even consulting her to see what her own preference might be. "There were two people [at Smith] doing research in algebra, and I would have loved to have gone to Smith for that reason," Schafer recalls ruefully. "The man who handled student placement at Chicago wrote back to Smith and said, 'No, she won't take that salary'" (Schafer: 18).

Why did the placement office turn down the opportunity at Smith without consulting Schafer first? Did they perhaps prefer to negotiate with a college where there was already a Chicago connection? There were no Chicago alumnae on the faculty at Smith, whereas Randolph-Macon had Gillie Larew and Evelyn Wiggin, and Connecticut College had Julia Bower. All three had earned their degrees under G.A. Bliss at Chicago, in 1916, 1936, and 1933, respectively.

"Julia Bower . . . was at the mathematics meetings, which were held in Chicago right about the time I was completing my Ph.D.," Schafer recalls. "Bliss pulled me over and said, 'I want you to meet Julia Bower. She was my student, and she may give you a job some time'" (Schafer: 18). Indeed, Schafer ultimately received offers from both Randolph-Macon and Connecticut, the outcome of a mysterious process of negotiation about which she knew nothing. Her experience was somewhat unusual, in that the placement system normally resulted in a single offer being extended to each woman candidate. She decided to take the job at Connecticut.

Sometimes job offers came through an informal network of connections that circumvented the Ph.D.-granting institution altogether. For example, fully a year before Winifred Asprey was to complete her Ph.D. at Iowa, Mary Evelyn Wells, the department head at Vassar, contacted her former honor student directly and invited her to join the Vassar faculty. Asprey wisely declined the offer, preferring to remain at Iowa to complete her Ph.D. Perhaps Asprey had reason to believe that the offer would be extended again, as indeed it was just one year later. At that point Asprey was all too happy to accept the opportunity to return to her alma mater as a member of the faculty. "So," Asprey recalls, "I had no job search" (Asprey: 25).

The old-boy/old-girl system continued on through the late forties and into the fifties, as evidenced by the fact that Grace Bates, Maria Steinberg, Augusta Schurrer, and Barbara Beechler also landed their first academic positions this way. Anne Calloway, who remained at the University of Pennsylvania as an instructor with research responsibilities for two years following the completion of her Ph.D., sought the assistance of J.R. Kline, the department head at Penn, to find a more permanent job in 1951. Although Calloway had been continuously involved in research and publication for several years, Kline's connections served only to produce job leads at women's colleges—first at Randolph-Macon Woman's College in Virginia and later at Goucher College in Baltimore, where she accepted an assistant professorship.

When a young woman was hired on to the faculty of one of the older women's colleges, the institution frequently had quite specific long-term career expectations in mind for her. There is no question, for example, that Mary Evelyn Wells hired Winifred Asprey to be her successor at Vassar.[4] When Marion Marsh Torrey (Ph.D., Cornell, 1924), the department head at Goucher College, hired Maria Steinberg (in 1948) and Anne Calloway (in 1951), she clearly had big plans for them. Calloway, in particular, was hired "with the understanding that I would learn enough astronomy to be able to teach the course and, eventually, probably take over as chairman of the [mathematics] department" (Calloway: 14).

In general, the older generation of women on the faculty of the women's colleges had spent their entire careers at one institution, building its reputation from the ground up. For the most part, these women had never married or had married late. In hiring fresh women Ph.D.'s in the forties and fifties, they were grooming a new generation of leadership at their institutions. The women of this younger generation who fit most comfortably on the staffs of the women's colleges tended to be those whose personal lives fit the mold of the older generation.

So, for example, Winifred Asprey *did,* in fact, serve as the successor to Mary Evelyn Wells on the Vassar College faculty. From the time she arrived in 1945 until her federally mandated retirement in 1982, Asprey was deeply engaged in the life and work of Vassar College. Even in retirement, she continues to live in a house on the campus and is vitally involved in college life. When asked if she has ever entertained serious thoughts of marriage, she replies: "I was so busy doing so many other things! It certainly wasn't a vital part of my life" (Asprey: 46). Grace Bates was similarly disinclined to marriage: "I always used to say, 'I'd like to have children, but I don't want a husband'" (Bates: 21). She lived on the Mount Holyoke campus and remained an active member of the faculty until her mandatory retirement at age sixty-five.

Anne Lewis Anderson served on the faculty of the Woman's College of the University of North Carolina (later, the University of North Carolina at Greensboro) for nearly twenty years and succeeded Helen Barton (Ph.D., Johns Hopkins, 1926) as head of the mathematics department in 1960. Although Anderson had very much wanted to marry, she eventually came to the belief that "I probably would not marry, that the opportunity wouldn't present itself." But in 1964 she met and married D.B. Anderson, vice president for academic affairs of the University of North Carolina system. Following the model of an older generation of women faculty, she resigned her teaching job to become his "full-time wife" and was not significantly involved in mathematical activity again afterward (Anderson-1: 8).

But for many other women of the younger generation of Ph.D.'s, the old career model of teaching at a women's college or state teachers college until marriage or retirement was badly mismatched to their preferences and talents. This mismatch was but a symptom of a much larger problem. Advisers didn't naturally see their female students as becoming full members of the mathematical community but rather saw them in a limited role: if unmarried, they would pursue a career in teaching; once married, their professional lives would essentially come to an end. Unlike their male counterparts, female doctoral students in mathematics were not generally perceived as bringing distinctive preferences and abilities to the job market. As a rule, if a woman Ph.D. wanted to achieve a career mix of research, teaching, and other mathematical and nonmathematical activities, it was up to her to devise the means of attaining a fuller role in the community. Furthermore, it was still a formidable challenge in the forties and fifties for a woman to combine a faculty position or other substantial professional activity with marriage. Thus it is not surprising that many of the women who were successfully placed into faculty positions at women's colleges in the forties and fifties were unable to conform to the expectations of the institution and often felt a mixture of dissatisfaction and alienation as a result.

So, for example, when Mary Dean Clement moved from the University of Chicago to Wells College on the eve of World War II, she experienced a kind of culture shock. Having spent her life in and around large cities—Nashville, Boston, Chicago—she was unprepared for life in "a minuscule college in a minuscule town" (Clement: 22). The winters were harsher than Chicago's had been, and once the war broke out, very few men were either on the campus or in the town. Moreover, Clement felt that she was being exploited by her (male) department head. "Although I'd had four years of teaching experience, he did not advance me beyond the rank of instructor for I don't know how long—I think three or four years," Clement recalls. Her salary was low, substantially lower than his because "he got an extra bonus in salary for having a wife." Moreover, she had "fifteen hours of teaching to his nine," and he taught nearly all of the upper-division classes (Clement: 23).

Indeed, Clement was so unhappy that, in a move unusual for its time, she investigated (to no avail) the possibility of teaching at Bryn Mawr or Vassar. Finally, during the 1946–47 academic year, she turned in desperation to her adviser, E.P. Lane, imploring him to "find me *something!*" (Clement: 25). In the thick of the postwar expansion, Lane helped Clement get a job at the the University of Miami, a rapidly expanding coeducational institution. The head of the mathematics department there was a Chicago graduate, Herman Meyer (Ph.D., Chicago, 1941), who was desperately seeking qualified personnel to add to his faculty.[5] At Lane's recommendation, Meyer hired Clement, who joined the Miami faculty in 1947 at the rank of instructor—a demotion from her rank at Wells, where she had been an assistant professor.

For Barbara Beechler, who had grown accustomed to the heady research environment in graduate school at Iowa and the "intellectual heaven" of teaching at Smith College, the transition to the secluded setting of Wilson College on receiving her Ph.D. in 1955 can only be described as traumatic (Beechler: 24). At Wilson, a small Presbyterian women's college in rural Chambersburg, Pennsylvania, Beechler taught four courses a semester. Her third year at Wilson saw the resignation of her lone colleague in mathematics, Roberta Johnson (Ph.D., Cornell, 1933). In 1958, just three years after receiving her Ph.D., Beechler was promoted to associate professor and appointed head of the department—an abrupt and overwhelming increase in professional responsibility.

At the same time, her position at Wilson afforded her little freedom in her personal life. She lived on the campus in a dormitory, with responsibility for the students in both dining hall and chapel. But there was clearly a double standard for male and female faculty. Faculty men were expected to have family lives, which they were free to keep somewhat separate from their work at the college, but women were expected to "[give their] all to the institution" (Beechler: 28). While there was no rule against faculty women marrying, single women were required to live on the campus, where living conditions were so claustrophobic as to effectively prevent the development of personal relationships that might lead to marriage.

Life on the campus at Wilson ultimately became unbearable. The last straw came for Beechler when her closest female friend on the faculty departed to take a position as dean at a prep school. In 1960 Beechler finally left Wilson for a position at Wheaton College in Massachusetts. Wheaton was less isolated, geographically and mathematically, than Wilson had been, but it was still a small women's college, where unmarried women faculty were expected to live on the campus and devote a substantial amount of their time and energy to college life. While Beechler quickly established herself as a successful and popular teacher and administrator at Wheaton, she felt increasingly suffocated within the women's college milieu.

Maria Steinberg is one woman of the forties generation of Ph.D.'s who launched her professional career at a women's college but ultimately moved on to a very different sort of career. With just a few revisions to make on her doctoral dissertation, Steinberg left graduate school at Cornell during the summer of 1948 and took a teaching job at Goucher College. At Goucher, she taught just nine hours a week, a far less onerous teaching load than that required at many other women's colleges (and an improvement on the eleven-hour load she'd had as a first-year graduate student at Cornell). There was plenty of time for her to work on her dissertation and take part in the social life of the college. But at the end of her first year there, Steinberg decided that she didn't particularly care to live in the city of Baltimore. Ph.D. in hand, as yet unmarried, and nearly thirty years of age, she was primed and ready to live in a place of her own choosing.

Taking matters into her own hands, Steinberg applied for and was awarded a Bateman Fellowship at Caltech in Pasadena, California. From 1949 through 1951, she worked as a research assistant on the Bateman project, which ultimately resulted in the publication of five handbooks of integral transforms and special functions. Her name appears in the list of Bateman project staff members at the front of each volume as the lone woman among the ten mathematical contributors.[6] Along with the prestige of her position at Caltech, she had the pleasure of living and working in California, which she quickly made her home.

Taking Charge, Taking Chances

For the women mathematics Ph.D.'s of the forties and fifties, there were essentially two accessible models of academic career development: the women's college model, which entailed nearly selfless devotion to the college and emphasized teaching, generally to the exclusion of research, and the emerging model of the research career provided by the myth of the mathematical life course. As we have seen, for many of the women of this younger generation, the women's college model was unattractive and incompatible with their professional talents and aspirations. At the same time, the model provided by the myth diverged sharply from the realities of women's lives and was nearly impossible to realize because research universities were frankly unprepared to accept very many women onto their faculties.

Many of the interviewees had serious research aspirations; some of them wanted to live in a particular part of the country; still others craved the experience of moving on to new challenges. Fulfilling these desires required them to take charge of their own professional destinies, to become trailblazers, willing to take the kind of risks that were necessary to break out of the prevailing model of mathematical careers for women.

As we saw in chapter 5, Dorothy Maharam Stone and Domina Spencer were two mathematics Ph.D.'s of the early 1940s who maintained, throughout their undergraduate and graduate school years, an intense and fiercely independent focus on their intellectual interests, a focus that sets them apart from many of the other interviewees. Maharam Stone was blessed with a perceptive and savvy adviser, Anna Pell Wheeler, who knew better than to send her student through the old-girl network to a women's college, where her research career might be imperiled. Instead, Wheeler used her influence to ensure that Maharam Stone was the recipient of the first Emmy Noether Fellowship awarded by Bryn Mawr College. The fellowship provided a stipend that enabled Maharam Stone to spend the academic year 1940–1941 at the Institute for Advanced Study in Princeton, concentrating on problems in measure theory.

During that year and the year following (during which the Institute subsidized her stay), she benefitted greatly from contact with other giants in the field, most notably Shizuo Kakutani and John von Neumann. At Princeton, she recalls, "I felt very much like a very small dog in very tall grass!" (Maharam Stone: 7). During her stay at the Institute, she produced two research papers; the second of these, representing a significant advance beyond what she had done in her dissertation, appeared in the prestigious *Proceedings of the National Academy of Sciences* in 1942. The mathematical contacts she enjoyed at Princeton helped to shape her research program for over fifty years thereafter (Oxtoby 1989).

Although many other women mathematicians had spent time at the Institute before Maharam Stone's arrival in 1940, few of them had been able to sustain their research momentum much beyond the limits of their stay. This was a problem as well for Maharam Stone. In 1942 she broke her childhood vow to remain single and married Arthur Stone (Ph.D.,

Princeton, 1941). Their marriage was followed immediately by a period of wandering, which included wartime stays at Purdue University and in Washington, D.C., and a move to England in 1945, where they settled in at the University of Manchester two years later.

During these peripatetic years, Maharam Stone gave birth to a son and a daughter and experienced a hiatus in publication. But she continued to think about mathematics, sustained by her own inner resolve, Arthur's ongoing professional employment, and the mathematical companionship she enjoyed with him. By 1947—the year that they settled in Manchester and the year that their daughter was born—her third publication appeared; a steady stream of publications followed thereafter.

Dorothy Maharam Stone did not hold a professional position of her own again until 1950, nor did she hold a regular professorship until she and Arthur joined the faculty of the University of Rochester in 1961. Yet her involvement with mathematics continued unabated during these eventful and unsettled years. Maharam Stone was keenly motivated to do mathematics regardless of her external circumstances, possessing an inner drive that enabled her to persevere without the external validation of a high-status professional position. It is difficult to imagine that very many men would have been able to sustain Maharam Stone's level of mathematical productivity under similar circumstances.

Domina Spencer also had a strong inner resolve to make mathematics and science her life's work. Unlike Maharam Stone, however, Spencer was determined from the outset to have the prestige and power of a professorship of her own. Moreover, she was intensely committed to her mother and sister, who had made many sacrifices for her education, and during her years at MIT she developed a personal and professional attachment to Parry Moon, an electrical engineering professor with eclectic interests in applied mathematics and the sciences.

Determined as she was to satisfy her personal and professional needs, Spencer took charge of her own job search. "I wanted . . . one of two things," Spencer recalls. "I wanted to be where Parry was; I wanted to be where my sister Vivian was" (Spencer: 24). This meant applying for jobs in the Boston and Washington areas. Spencer cast her employment net as widely as possible. She had long viewed herself as an interdisciplinary scientist, and so with her customary confidence, she marketed herself as such, seeking jobs in both physics and mathematics. It was an audacious strategy for a woman in the turbulent 1940s—but ultimately, it was a strategy that worked.

The summer of 1942 was a particularly auspicious time for Spencer to seek faculty positions in mathematics and physics. "Science faculties were badly depleted," writes the historian Margaret Rossiter, "when many men (who had held most of the academic jobs in the 1930s) left suddenly for governmental science projects or, if young enough, the military" (1995: 9). Despite a steep drop in enrollments, science courses were more popular than ever before. Thus it is not surprising that three universities expressed interest in the twenty-one-year-old Spencer.

"I got an interview with Northeastern," she recalls, but when she came to the campus it was clear that despite their staffing needs they were not ready to hire a woman. "They told me I wouldn't be happy alone there because they had no other women," Spencer says; they feared she would be "lonesome." But after her experience at MIT, Spencer was thoroughly accustomed to being "the lone woman among a large group of men, and I didn't think anything of it. . . . They didn't make me an offer, they claimed, because I wouldn't be happy there" (Spencer: 25). Ultimately, she did receive two job offers: an instructorship at the University of Maryland at a salary of $2,400 a year and, at the same salary, a position as "assistant professor and chairman of the physics department" at American University. Urged by her sister to "take the school that gives you the higher rank," Spencer took up her duties at American University in the fall of 1942 (Spencer: 25).

The security of her new job, however, was in question from the very start. American University was "not a terribly strong school in science. It had one or two professors in each area. And that wasn't enough" to secure an army or navy training program (ASTP or V-12) on the campus in 1943. Facing an impending enrollment crisis, the president of the university, Paul Douglass, called a general meeting of the faculty in the spring of 1943 and advised them to seek jobs elsewhere.[7] After less than a year on the job, Spencer recalls, "I sent out another batch of applications." This second round on the job market sent her back to the Boston area where, in the summer of 1943, she was hired as an assistant professor of physics at Tufts College. The president of the college, Leonard

Carmichael, informed her, "If you do well, we may have another woman some day" (Spencer: 25–26).

Spencer stayed on the faculty at Tufts for just four years. She attributes her dismissal in 1947 to the fact that she got on the wrong side of a political dispute in the physics department; but in the early postwar years, women who had been hired onto college and university faculties during wartime were regularly dismissed to make room for increasingly available men.[8] Still, Spencer made the most of her years at Tufts. She was in close proximity to Parry Moon, who kept an extra desk in his office for her at MIT. Working alone and with Moon, Spencer worked to build up her research reputation in physics, engineering, and mathematics.

In 1947, Spencer submitted a third round of job applications, this time receiving expressions of interest from two Ivy League schools, Harvard and Brown. She recalls that an offer from Harvard's engineering school was extended to her but withdrawn less than two weeks later, with an excuse that only thinly veiled Harvard's reluctance to hire a woman into a regular faculty position. Brown's interest in her was apparently more genuine, and she was hired into her third assistant professorship in physics on a three-year contract.

Her years at Brown, though happy ones, were problematic. Research faculty were expected to do research under government contracts, but her Quaker sensibilities prevented her from doing so. "I refused to work on any government contract that was sponsored by a military organization," she recalls. "If I had not done that, I could have stayed at Brown forever" (Spencer: 30). As her initial appointment was about to expire in 1950, her future at Brown was in doubt, and she carefully pondered her next move.

In the eight years since she had earned the Ph.D., Spencer had established a reputation as a first-rate researcher and had particularly distinguished herself in the up-and-coming field of illuminating engineering. In the summer of 1950, as she wearily contemplated submitting a fourth round of job applications, her hard work finally paid off. Gregory Timoshenko, an electrical engineer and a collaborator of Moon's, and W.F. Cheney, Jr. (Ph.D., MIT, 1927), the head of the mathematics department, aggressively recruited her for a position in mathematics at the University of Connecticut. They eventually won her over by offering her a teaching schedule that would enable her to commute comfortably from her home in Cambridge two or three days a week. She joined the faculty as an assistant professor, was promoted to associate professor with tenure the following year, and was made a full professor in 1960. She continues on the U. Conn. faculty to this day, commuting from the Boston area to the campus in Storrs just as she has done for nearly fifty years.

Both Maharam Stone and Spencer shared a clear and unwavering commitment to creative mathematical work. While they were assisted by friends and faculty who advocated for them and wrote letters of support, they never *relied* on those contacts, as participants in the old-girl/old-boy network were encouraged to do. Their mathematical activity was simply too important to them to be left in the hands of others. Spencer's determination, in particular, to remain close to family and to Moon, to adhere to her pacifist principles, and to have the collegial respect of her scientific peers meant that she had to return to the job market again and again, enduring not one but *four* assistant professorships. In the end, her perseverance was rewarded when she found a department seeking not merely to fill a position but to fill it with *her*.

Breaking the Mold at the Women's Colleges

Among the women who landed their jobs the old-fashioned way and made their living in the women's colleges, there were certainly those who took control of their destiny—exploiting the modicum of independence their academic lifestyles gave them, taking advantage of unique opportunities in the postwar era, and exploring new avenues of personal and professional development. Winifred Asprey's long affiliation with Vassar College provides perhaps the best example of this among the interviewees.

Before she got the job offer from Vassar, Asprey entertained thoughts of teaching college in Alaska. "I'd never been to Alaska before," she recalls. "I thought it would be a lot of fun to do." To his credit, her dissertation adviser, E.W. Chittenden, did not try to talk her out of it; rather, when the offer from Vassar came, he persuaded her that "it would be much more prestigious, both for me and for him, if I went back to Vassar" (Asprey: 25).

Asprey took Chittenden's advice, thinking that she was only temporarily postponing her Alaskan adventure. "I knew that Miss Wells was going to retire in three more years," Asprey recalls, "so I thought, 'Well, I'll go to Vassar for three years, and then I'll take off'" (Asprey: 25). Once at Vassar, however, dreams of Alaska began to fade as she found herself very much caught up in the intellectual, political, and social life of the campus. Promotions—from instructor to assistant professor to associate professor to professor—and appointment as department chair came in a timely fashion. Like Barbara Beechler, she was expected to live on the campus and for a time supervised a dormitory. But the Vassar community provided sufficient variety and diversity that Asprey did not experience the same isolation and suffocation that Beechler did.

Even so, after ten years at Vassar, Asprey became restless. "A very wise friend pointed out to me that I was in a rut," she recalls. "I had been chairman of the department; I'd been on the most important committees. . . . I was lecturing all over the country and even abroad. Was I simply going to repeat these experiences until I retired?" (Asprey: 44). The key to renewing her enthusiasm was to strike out in a new direction.

She began by calling on her old Vassar professor, Grace Murray Hopper, who was by this time working with UNIVAC in Philadelphia. "I went down and spent four perfectly marvelous days," says Asprey. "She was teaching the machine differential calculus at the time. Then she said, 'How would you feel about coming down here, and spending a year, and seeing what you can learn about'—she called it 'this beast'—'this beast?'" Hopper's offer was terrifically tempting: "We were very good friends; I liked her so much, and she was so inspirational: teaching, lecturing, every possible way of friendship" (Asprey: 32). After considerable soul-searching, Asprey declined Hopper's offer, choosing instead to blaze a trail into computing that gave her a sense of independent accomplishment and at the same time reaped important benefits for Vassar College.

In the late fifties and early sixties, the research laboratories of IBM were located near Vassar in Poughkeepsie, New York. Summoning "all the courage I ever had," Asprey contacted IBM and offered them her services as "a pure mathematician" (Asprey: 33). As it happened, IBM was interested in working with Asprey and awarded her one of their first industrial research postdocs. At an early stage in what was to be a long association with the company, IBM offered Asprey a permanent job at "an astronomical salary." The temptation was great, just as it had been when Hopper invited her to UNIVAC. But once again, Asprey made the choice that would maintain her independence, which she says was "the thing I treasured most at Vassar. . . . At IBM, you had to ask every step of the way; you had to do what your manager directed you to do. I was more than willing to settle for the lower salary and my independence" (Asprey: 45).

In the years that followed, Asprey went from a position of near-total ignorance to an advanced level of computer expertise. Over time, she worked to establish a cooperative partnership between IBM and Vassar, in which IBM employees came to the campus and taught computing to Vassar students. The culmination of over ten years' work in computing came during the academic year 1966–67, with the installation of a "state-of-the-art, IBM 360 Model E, with memory capacity of thirty-two K," in the brand-new Vassar computer center, over which she presided as founding director (Asprey: 38).

The decision to remain at Vassar while working closely with IBM reinvigorated Winifred Asprey. She established a vital link between the comparatively isolated world of the women's college and the interdisciplinary, business-oriented world of computing. At a personal level, Asprey's decision empowered her to shape her own environment and gave her access to a far wider circle of social and professional contacts than she had ever had before. Over the years, Asprey—restless, curious, gregarious, and indefatigable—broke the traditional mold of the women's college mathematics professor, creating the kind of life and career she wanted and needed.

New Opportunities in the Mathematical Sciences

Whether of necessity or by design, several women of the forties and fifties generation opted out of both of the prevailing models of the mathematical career and took a completely different professional route. Margaret Martin, who wrote a dissertation on orthogonal polynomial systems at the University of Minnesota in the early forties, became disenchanted with her research subject at a fairly early stage. "I couldn't see where it was leading, and in fact I didn't come up with what my adviser anticipated I would," she recalls. "And whether that was because I failed to do it or because it wasn't there, I wasn't sure. But it was rather esoteric, and it wasn't something I particularly enjoyed, so that I did it because I wanted to get my degree" (Martin: 10).

What redeemed the graduate school experience for her was the enjoyment she took from her minor subject, statistics, which she says she "liked because it dealt with the real world" (Martin: 8). There was no question in her mind as to the comparative value of mathematics and statistics:

I thought of mathematics as a kind of fossil subject. All of the interesting things had been done in the nineteenth century, I thought. And, you know, with the limitations that I had, I didn't see that one had much of a chance of developing anything that was really important. You could do research, but it wouldn't have much practical application. I guess I thought just about everything in math that was worth doing had already been done. And statistics was this new, developing subject, which had all sorts of applications in the real world, and it was a lot more attractive. (Martin: 11)

Martin came to view the Ph.D. in mathematics, in essence, as a credential that—when combined with her minor in statistics—would open doors for her in the statistical world.

Statistics was such a comparatively new field in the early 1940s that there were very few doctoral programs and no clearly established career path for those with graduate training in statistics.[9] As soon as her coursework for the Ph.D. in mathematics was complete, Martin, "tired of living on a shoestring," sought and obtained an instructorship in statistics at the University of Minnesota for the academic year 1940–41 (Martin: 11). In 1941, she moved to New York City to take a job with a testing service but in fairly short order moved on to an instructorship in biostatistics at the Columbia University School of Public Health.

She spent the years 1942 to 1946 in a teaching position at Columbia, which was as much an apprenticeship in statistics for her as it was for the students she served. Her students were, for the most part, "physicians—M.D.'s—who were working in health departments" and had come back to earn master's degrees in public health. "My boss, John Fertig, did the teaching, and I assisted him and worked with students, answering their questions and helping in various ways," Martin recalls. "I really liked that job because my boss was a live wire. He taught me *so* much about statistics!" (Martin: 14).

Working on her dissertation mainly during vacations, Martin finally completed her Ph.D. in 1944. By that time, however, she was clearly committed to a future in statistics. In 1945, while still at Columbia, Martin worked as a statistical consultant for the New York City Health Department. Her years of experience in New York City—which she left, in 1946, for health reasons—served as a springboard to professorships in applied statistics—first at the Vanderbilt University Medical School and later at the Johns Hopkins University School of Public Health.

Joan Rosenblatt, coming of age a decade later than Margaret Martin and with the benefit of her parents' experience as college professors, decided at a much earlier stage that she preferred statistics to mathematics, and "real-world" applications to strictly theoretical work. During the ten years that elapsed between the completion of her bachelor's degree at Barnard College in 1946 and the completion of her Ph.D. in mathematical statistics at the University of North Carolina in 1956, she had short-term employment at several federal agencies, including the Bureau of the Budget and the Census Bureau in Washington, D.C. These experiences confirmed her desire to do applied work in the public sector. What is more, they put her in contact with people who could hire her on completion of the Ph.D. One of these people was Churchill Eisenhart, director of the statistical section of the National Bureau of Standards, who did, indeed, hire her as a mathematician in the Statistical Engineering Laboratory of NBS in 1956.[10]

Margaret Martin and Joan Rosenblatt enjoyed the freedom to choose career paths that were somewhat unconventional for women mathematicians of the time and that led them into exciting new areas of applied mathematical endeavor. Evelyn Granville similarly chose an unconventional career path that led her into new fields of mathematical application, but for Granville, the choice was constrained by racial discrimination. Granville received the Ph.D. from Yale in 1949 and spent the following year in a research postdoc at NYU, working on a project in differential equations under Fritz John. While she enjoyed the experience of "being associated with these really high-level mathematicians" at NYU, the time she spent there was not particularly productive in terms of research. In particular, she recalls that her Yale adviser, Einar Hille, never encouraged her to publish her dissertation; nor did any publications under her name result from her association with John (Granville: 10–11). It is not clear what constructive guidance, if any, either of these men offered her as she set out on her mathematical career.

As her year at NYU drew to a close, Granville decided to seek a college teaching job. She was granted two interviews. Her first interview, at the Polytechnic Institute of Brooklyn, seemed to go smoothly, but she was not offered a position there. Some time later, she "was told that they thought it was a big joke that a colored woman would apply there for a teaching job." They probably hadn't realized that she was black until she came to campus for her interview, and in 1950, Brooklyn Polytech was apparently not ready to have a black woman on its faculty. Her second interview, with the president of historically black Fisk University in Nashville, was held in New York City and resulted in the offer of an associate professorship to begin in the fall of 1950. It was a position she was happy to accept; by this time, she had decided that New York City was "not my cup of tea" (Granville: 11).

Nashville was a segregated Southern city, which allowed a young woman of color very little freedom of movement or association. "We confined our life to the campus," Granville says. "Our whole life was right there on the campus. . . . In a sense, we were in a little island, and so the city didn't bother us at all." But the segregation of Nashville posed problems for the faculty at Fisk nonetheless. "Once there was going to be a math meeting at Vanderbilt," Granville recalls. "We wanted to go; we were going to attend the math meeting and also attend the banquet." Lee Lorch, the white professor and civil rights advocate who chaired the Fisk mathematics department at the time, informed his faculty that the meeting's organizers "did not want us—at least they didn't want *black* faculty—to attend the banquet."[11] Granville took the news in stride: "I don't recall being terribly upset about it, maybe because I had grown up in a segregated environment. You were just accustomed to accepting things . . . as they were" (Granville: 12).

In the final analysis, however, Granville—who had wanted to become a teacher since she was a little girl—was unable to accept the highly restrictive terms under which black women could hold academic positions in the early 1950s. As she considered her options, it was natural for her to think about the possibility of government employment; after all, the federal government had been the most liberal employer of black women during World War II.[12] In the spring of 1952, Granville decided to seek a government job and return to Washington, D.C.

"I remember a young man, a black fellow, head of a group at the National Bureau of Standards, came to Nashville to interview me for a job," Granville says. His offer to double her salary, "from something like thirty-six hundred dollars to seventy-two hundred dollars a year, was not something to be sneezed at. And so I came back to Washington" (Granville: 14). From 1952 until 1967, Granville worked at a succession of jobs in government and industry—first in Washington, briefly in New York City, and (from 1960 onward) in Los Angeles. Over the course of fifteen years, she developed expertise in computer programming and was intimately involved in research and development projects for the space program. Discrimination had led her away from the teaching career she had longed for, yet she was able to transform her misfortune into the opportunity to get involved in two exciting new fields—computing and aerospace—during their infancy.

The unconventional career paths followed by Martin, Rosenblatt, and Granville were largely compatible with their personal lives. Martin never married; Granville married for the first time in 1960. During the crucial early years of their careers, both had the freedom and the mobility to pursue professional opportunities as they arose. Joan Rosenblatt *was* married, but she and her husband chose not to have children and shared the sense that Washington was the professional center of gravity for *both* of their careers.

Marriage and Its Complications

However, many women mathematics Ph.D.'s of the forties and (especially) the fifties married and had children at an early age. Their family responsibilities sharply limited their freedom to explore the career options available to them. Moreover, in a cultural period that valorized the domestic role of women, many of them suffered the effects of discrimination—both in the workplace and in the mathematical community—when they tried to secure the kind of employment for which their Ph.D.'s had prepared them. Those women who succeeded in maintaining an active and fulfilling professional life generally did so by accepting certain limitations, challenging others, and, in general, taking an inventive and open-minded stance toward their employment options.

In stark contradiction to the myth, a surprising number of these women were able to return to a more intense, and frequently more creative, engagement with mathematics *after* they had devoted several years in their twenties and thirties to the rearing of children. Women who returned to active mathematical careers in the 1960s and 1970s sometimes enjoyed an unanticipated benefit: the litigation and legislation of the women's movement opened many doors that had been closed to them earlier, as increasing numbers of coeducational colleges and universities welcomed women to their faculties.

A sense of equality in marriage, along with a strong sense of professional identity, may have been easier to achieve for those women who did not have children. Violet Larney and Herta Freitag provide the best examples of this. As Violet Larney was finishing her doctorate at Wisconsin, she met and became engaged to Norbert Larney, who was at the same time completing his *undergraduate* degree in electrical engineering. They were approximately the same age, Norbert Larney having served in the Royal Canadian Air Force during World War II. When the time came for Violet Larney to look for a job, her adviser, C.C. MacDuffee, simply advised her to wait and see where Norbert got a job and then try to find a position in the same place. Larney was aghast: "I don't want to be just a substitute teacher or something in some place just because my husband's there!" she told McDuffee. "I worked *hard* all these years to get three college degrees in mathematics!" (Larney: 12).

Indeed, when Larney got her first job at Kansas State College in 1950, her fiancé was the one who who tried to get a job nearby. Unfortunately, the closest engineering job that he could find was in Chicago. When they were wed in Chicago that fall, they rapidly discovered that the strain of a commuter marriage was more than they could handle over the long haul. Over the course of the next two years, Violet and Norbert Larney took turns making professional and geographic moves until both of them landed in the same metropolitan area in jobs that they liked. Norbert Larney made the first move, spending the year 1951 pursuing a master's degree on the G.I. Bill at Kansas State. In 1952 they once again separated, as Norbert took a job with General Electric in Massachusetts.

At that point, Violet Larney appealed to MacDuffee for help. MacDuffee contacted his former Ph.D. student, Caroline Lester (Ph.D., Wisconsin, 1937), who was on the faculty of the New York State College for Teachers in Albany, just forty miles from the General Electric plant. The college hired Violet Larney as an associate professor to begin in the fall of 1952. At the same time, Norbert Larney sought and obtained a transfer to General Electric's plant in Schenectady, closing the distance between home and work still further. Once they were finally settled in Albany, Violet and Norbert Larney—who were thirty years old when they married—made their careers the main focus of their lives. Working in separate fields and not having children helped them to feel that their marriage was a partnership of equals.

When Herta and Arthur Freitag married in 1950, Herta was nearly forty-two years old and Arthur was ten years older; it was clear to both of them at the outset that they would not have children. At the point at which they decided to marry, Herta Freitag was deeply immersed in her teaching duties at Hollins College near Roanoke, Virginia, and working on her Columbia Ph.D.; Arthur Freitag was a high school administrator in Chicago. As Herta Freitag remembers it,

He came, and he said, "I have been thinking things over. You love it at Hollins. To me, a job is a job. Why don't I pull up stakes and come to you? As a Yankee, I won't get into educational administration down there, but I'll get a teaching position, and that will be good enough for me." (Freitag-2: 17)

Arthur Freitag was decidedly not a young man just setting out on a career when he married; just the same, it was an extraordinary sacrifice for a man of his time to subordinate his own career to his wife's. Moreover, throughout their marriage, he did everything possible to support Herta in her intellectual and professional endeavors. Herta Freitag reports that her husband "felt that I, being a college teacher, would have very much more work to do than he. So he just set himself to take everything away from me that he possibly could." In particular, Hollins College held classes on Saturday, so Arthur Freitag "used that Saturday to do the cleaning, the washing, the ironing, the grocery shopping, and everything else" (Freitag-3: 15).

Both the Larneys and the Freitags were older than average when they married. In each couple, at least one partner was making up for time that had been lost because of World War II. Each couple was highly motivated to combine satisfying work and a warmly companionable marriage. The satisfaction that they felt, both at work and at home, was greatly enhanced by the fact that neither partner had extraordinary career ambitions, but each was content to make their mark in relatively unpretentious circumstances. Both couples enjoyed an extraordinary compatibility of values and goals.

Many of these same qualities can be found in the marriage of David and Joan Rosenblatt, who, like the Larneys and Freitags, also did not have children; and in the unusual partnership of Domina Spencer and Parry Moon. Spencer and Moon had been companions and partners in research for fourteen years when, in 1956, Moon's estranged wife died of a heart attack. Five years later, when Spencer was forty and Moon sixty-two, they were married. By this time, Moon had been on the faculty at MIT for nearly forty years, and Spencer was a full professor at Connecticut; both of them lived in Cambridge and their personal and professional lives were fully intertwined.

In 1965, when Spencer was forty-four years old and her professional reputation was well established, she gave birth to a son. With the full cooperation of the close-knit Spencer clan, the upbringing of young Euclid Moon brought minimal disruption to Spencer's professional life. While Spencer taught her classes at Connecticut, her mother watched the baby in her office. "[W]hen I came back," Spencer recalls, "ten minutes between class I nursed the baby, and then I went to the next class. . . . When he got bigger, we had some of the students come and teach him languages" (Spencer: 33). In the Spencer family tradition, much of Euclid's early education was conducted at home, with the assistance of his aunt Vivian, who moved to the Boston area in the last years of her life.

The combination of Spencer's family circumstances—including the tremendous personal resources of her mother and sister and the extraordinary cohesiveness the three of them enjoyed—with her own pluck and good fortune can only be characterized as highly unusual. Other women of her generation encountered far greater logistical and professional difficulties as they tried to combine mathematics with marriage and family. In particular, nearly every woman who married a fellow mathematician, scientist, or other academic faced the need to subordinate her career to her husband's, at least in the early going. In marriages with children, the greatest challenge was to maintain some level of mathematical activity and momentum during the early years of pregnancy, childbirth, and child rearing. In many of these marriages, the women eventually achieved a kind of professional parity with their husbands, usually after the children had grown. Some marriages, of course, cracked under the strain.

During the earliest years of their marriage, Dorothy Maharam Stone followed her husband Arthur as he moved through a sequence of professional positions. But in 1961, at a time when antinepotism rules were in force at many colleges and universities in the United States, the Stones were hired together as full professors in the mathematics department at Rochester. Since that time they have enjoyed a professional equality that has been unusual for academic couples until quite recently. As Maharam Stone's former Bryn Mawr professor, John Oxtoby, once observed, "There have been other husband-wife teams in mathematics, but probably none in which each has so consistently stimulated and participated in each other's work" (1989: xviii).

However, it was not uncommon in mathematical marriages for the husband to have the research career while the wife did not. Both Maria Steinberg and Anneli Lax gradually stopped doing research after their marriages to fellow mathematicians. In both cases, the movement away from research appears to have been a conscious choice guided by their natural preferences and predilections, but it is impossible to know all the factors that entered into the decision in either case.

Maria Steinberg worked on the Bateman project at Caltech until 1951. "I drifted along after that," she says. "I wasn't really very self-motivated, as far as research is concerned" (Steinberg: 12). She had, however, fallen in love with California, and with a view to staying in her adopted home state, she took a two-year acting assistant professorship at UCLA. At UCLA, she met and married her husband, the mathematician Robert Steinberg (Ph.D., Toronto, 1948) and became a naturalized citizen of the United States. While she continued to hold professional positions in mathematics (Hughes Aircraft; Cal State Northridge), she did no further publishable research.

Steinberg believes that, in a two-career couple, "it is a good idea to let one person fashion the life" by having their career take precedence. In relation to her husband, she says, "I took second spot because I do realize he's a far better mathematician" (Steinberg: 18). But how does a couple decide which one is "better"? In the case of the Steinbergs, it seemed to be a matter of motivation. Although Steinberg's early research was of consistently high quality, she simply was not *driven* to continue in the same way that her husband was.

Anneli and Peter Lax met and married while both were graduate students at NYU. Events followed rapidly on their marriage in 1948: in 1949 Peter Lax completed his Ph.D. and joined the faculty at NYU; in 1950 and 1954 Anneli gave birth to two sons; in 1955 she finally received her own Ph.D. By this time, Anneli Lax had been a part of the Courant "family" at NYU for some twelve years, and Peter's own distinguished career in mathematical research was well under way. While Anneli continued her affiliation with NYU, her professional center of gravity shifted from research and graduate studies to the undergraduate program at NYU's Washington Square College, where she was awarded tenure in 1961.

Perhaps, during her long association with various projects at Courant Institute, Lax never fully experienced the sense of independent achievement that bolsters self-confidence and gives momentum to research. Despite her considerable early promise, Lax gradually drifted away from research mathematics. Over time, she came to be increasingly involved in teaching, expository writing, and editing, where she could make her own distinctive contribution.

Both Lax and Steinberg enjoyed a high degree of geographical stability in their personal and professional lives, but other women of their generation were not so fortunate. Perhaps the most extreme case among the interviewees who followed a husband's career is that of Alice Schafer. From the time they first met in graduate school at the University of Chicago, mathematics has formed the cornerstone of Alice Schafer's relationship with her husband, Dick. But there was no question that Dick's career moves would take precedence over hers, and from the outset, Alice Schafer's employment decisions were made as a counterpoint to his.

During the first two decades after Alice Schafer earned her Ph.D. at Chicago, she held a dizzying array of professional positions. She left her first job at Connecticut College after just two years because her husband was soon to be discharged from the navy and she wanted to be with him. After a brief stint at the Applied Physics Laboratory at Johns Hopkins University, she taught first at Michigan, where Dick held an instructorship from 1945 to 1946 and then at the New Jersey College for Women (later, Douglass College, Rutgers University) from 1946 to 1948 while Dick was at the Institute. From there, the Schafers moved on to the Philadelphia area, where Alice was assistant professor, first at Swarthmore (1948–1951) and later at Drexel (1951–1953), while Dick was on the faculty at Penn. During their years in Philadelphia, the Schafers adopted two sons—John in 1948, Richard in 1951—and from then on they hired a succession of women to help care for the children at home on weekdays.

When Dick moved to the University of Connecticut as department chair in 1953, Alice spent a year teaching part-time at U. Conn. before being rehired by Connecticut College, where she had begun her career eleven years earlier. This time she stayed on the faculty for nine years, achieving tenure and promotion to full professor. It was a blissful period of professional stability for Alice Schafer, but it was not without complications. In 1958, Dick moved on to MIT; this time Alice, recently tenured, did *not* move. In 1960, however, the Schafers bought a house in Belmont, Massachusetts, and from that time forward Alice Schafer taught in Connecticut during the week and commuted to Belmont on the weekends.

It was not until 1962, by which time Alice Schafer had had her Ph.D. for twenty years, that she received her first unsolicited offer of academic employment: Wellesley College, in the Boston suburbs, sought to replace the retiring Marion Stark (Ph.D., Chicago, 1926), and colleagues of Dick's from MIT and the Institute recommended Alice for the position. She went to Wellesley as a full professor in 1962 and at long last found respite from what had been an extraordinarily peripatetic academic career.

"Having It All"

Many of the other interviewees experienced similar kinds of disruptions in the continuity of their employment, but Schafer's experience is unique in the frequency and sheer number of job changes she experienced in a relatively short time. For this reason alone, it is not surprising that she never really got a research career off the ground but instead devoted most of her professional energies to teaching, administration, and public service. But other women—Jane Cronin Scanlon, Mary Ellen Rudin, Cathleen Morawetz, Jean Rubin, Edith Luchins, Patricia Eberlein—gave precedence to their husbands' careers and the needs of the children and yet were able to maintain a reasonable level of scholarly productivity. Cronin Scanlon, Rudin, and Morawetz were perhaps the most spectacularly successful of these in terms of the quantity and quality of research they produced over the course of their careers.

There are striking similarities and differences in the life and work patterns of these three women. All three were blessed with dissertation advisers who introduced them to fertile research areas. All three got their research programs off to a running start after graduate school, spending the years immediately following the completion of their Ph.D.'s in stimulating mathematical environments. Both Morawetz and Cronin Scanlon achieved tenure in a university mathematics department within ten years of receiving their doctorates, but Rudin never held a regular tenure-track faculty position until she was made a tenured full professor at the University of Wisconsin in 1970.

All three were married at an early stage in their careers—Morawetz, in 1945, to the chemist Herbert Morawetz; Rudin, in 1953, to the mathematician Walter Rudin; and Cronin Scanlon, also in 1953, to the physicist Joseph Scanlon.[13] Each woman had four children and had primary responsibility for child care. All three of them developed the ability to take small snatches of time for work and to tolerate a rather high level of distraction. Perhaps needless to say, they shared an intense inner drive to do creative work in mathematics. While Cronin Scanlon's family truly needed the money she earned from her professional employment, neither Rudin nor Morawetz had a strong financial incentive to work outside the home. In fact, Rudin says, "I spent more money than most on child care. It would have been cheaper for me to stay home" (Rudin: 18).

How did these women get their mathematical careers and their married lives off the ground? As is mentioned in chapter 5, Morawetz was married and had two children even *before* completing the doctorate. While still a graduate student, she knew that her husband's job would take priority over hers. "He was the money earner," she recalls; but he had also promised "that he would never do anything that would interfere with my career" (Morawetz: 12). Indeed, all of his major career moves were discussed and evaluated in advance in terms of their consequences, pro and con, for both Cathleen and the children.

Morawetz completed her dissertation research in the late summer of 1950 and moved with her husband to Cambridge, Massachusetts, where he had accepted a one-year fellowship at Harvard. She spent the year as a research associate in applied mathematics at MIT, where she had done her master's degree several years before. As a graduate student at NYU, she had the tendency to get stuck on intractible problems; at MIT, her mentor, C.C. Lin, would give her a problem to work on and would never let her keep at it for more than three weeks. If she had made no progress in that amount of time, he would set her to work on a different, perhaps related problem. In this way, she learned a valuable technique for coming "unstuck" and developed a sense that headway in research was possible. At MIT, she completed the work that led to her first two publications in mathematics.

A wiser and more self-confident Morawetz returned to New York and rejoined the research staff at NYU the following year, when her husband received an offer to join the faculty of Brooklyn Polytechnic. The move back to New York turned out to be permanent: Herbert Morawetz's career flourished at Brooklyn Poly, while Cathleen Morawetz eventually landed a tenure-track position at NYU and rose quickly through the ranks, becoming a full professor in 1966. In an institution where women seldom rose to positions of power or influence, Cathleen Morawetz was appointed associate director of the Courant Institute in 1978 and director in 1984. In 1995, she became the second woman elected president of the AMS.

Mary Ellen Rudin and Jane Cronin Scanlon, who share an uncanny synchronicity in the timetables of their early adulthood, each had four years of postdoctoral experience before they were married and another year before the birth of their first child. During these early years post-Ph.D., Rudin and Cronin Scanlon established contacts in the mathematical community, got research programs underway, and developed the momentum which kept them going mathematically through their childbearing, child-rearing years.

Rudin was hired as an instructor at Duke University in 1949; R.L. Moore, operating through his own personal old-boy network, arranged the position for her. Although she was required to maintain an affiliation with the women's college at Duke, Rudin's instructorship was a regular faculty position in the Duke mathematics department, where she benefitted from the mentorship of another Moore student, John H. Roberts (Ph.D., Texas, 1929). During her years at Duke, she met the legendary Paul Erdös and began to do "a lot of work with strange connected sets in the plane. . . . He had a whole series of such questions that he was interested in seeing solved, and I solved a number of them."[14] She remembers those years as "fantastically wonderful" and says that her mathematical work was and is so terrifically enjoyable that "it still seems impossible that anyone would pay me for doing this" (Rudin: 11).

During her first year at Duke, Mary Ellen met and dated Walter Rudin (Ph.D., Duke, 1949). She recalls that during that year, first one and then the other contemplated marriage, but "never was there a time *both* of us wanted to get married to each other" (Rudin: 12). After that first year, Walter moved on to a two-year position at MIT, followed by a tenure-track job at the University of Rochester, while Mary Ellen continued at Duke. When, in the spring of 1953, they at last decided to marry, R.L. Wilder had just arranged a visiting research position for her at Michigan, and Samuel Eilenberg had tried to recruit her to the faculty at Columbia. It was at this auspicious juncture that Mary Ellen Rudin resigned her position at Duke and moved to Rochester.

At Rochester, Rudin says, "I had two babies almost immediately." She taught courses on an as-needed basis but did not have a regular faculty position. "There was no one in Rochester in my field," she recalls. Thus it was up to her to "[find] things that I liked to do mathematically" and to attend as many conferences as possible so that she could find people with whom to discuss her research. When the children were young, she simply took them along; when they got older, she found someone to stay with them while she was away (Rudin: 14).

"Wherever I was, I went to look for the mathematicians," she recalls, "and I did whatever mathematicians were doing" (Rudin: 15). So, for example, when Walter spent a sabbatical at Yale in 1958–59, Rudin found that she had common interests with Shizuo Kakutani, and went into the department regularly to talk mathematics with him. When Walter moved to Wisconsin in 1959, she was pleased to find "really good seminars in topology" there, and the department gave her an office next to a fellow Moore student, RH Bing (Ph.D., Texas, 1945). She continued to teach part-time, to apply for and be awarded NSF grants, and in general, to maintain a steady focus on mathematics despite a constant level of distraction:

Even before any of my children were born, I worked at home, usually sitting down on the sofa. I liked to be comfortable, and I didn't mind noise particularly. And when [the children] were little, it was actually something very nice to do, to sit down with the little kids and let them do their thing while you sat and thought about mathematics in the same place. And you don't have to pay your entire attention to the child; the child goes about his business, just like you do. And so, actually, I didn't find this to be a problem for me, either with the first or the second or the third or the fourth. . . . If you decide in advance that you can't possibly work when there's some sort of distraction, then you don't work. Because there's something going on, off and on, all the time. But if you decide, "Okay, well, I'll think about this when I have a few minutes," then you think about it for a few minutes, you go do something else, you come back. Actually, that's how one does mathematics a great deal of the time. It's not with continuous thought. You don't sit down and solve a problem in some finite number of hours without interruption. You work at it for a little while; you go away from it a little bit; you come back to it. (Rudin: 16)

From 1959 until 1970, Rudin held the title of lecturer in the mathematics department at Wisconsin and was paid by the course. It was an arrangement that did not displease her: "I really had the best of all possible worlds. I wasn't on any committees. I taught what I wanted to, when I wanted to, the amount I wanted to, and I had four children" (Rudin: 15). Thus it came as something of a surprise when, in 1970, the mathematics department offered her a tenured full professorship.

During the late sixties, colleges and universities experienced steadily mounting pressure to show that they had qualified women in tenured positions and to demonstrate that they did not discriminate against women on the basis of their marital status. In 1967, President Lyndon Johnson signed Executive Order 11375, prohibiting sexual discrimination by federal contractors. In January of 1970, the psychologist Bernice Sandler joined with the Women's Equity Action League (WEAL), a nonprofit organization devoted to economic justice for women, in filing a complaint with the U.S. Department of Labor, charging the University of Maryland with sex discrimination and noncompliance with Executive Order 11375. The complaint was eventually expanded into a class-action suit against literally hundreds of institutions of higher education, including the University of Wisconsin. At about the same time, the American Association of University Professors began the preparation of a report, issued in the spring of 1971, that condemned antinepotism rules in college and university employment. By the summer of 1972, the struggle for gender equity in higher education culminated in the enactment of Title IX, which explicitly prohibited gender discrimination in colleges and universities receiving federal funding. The timing of Mary Ellen Rudin's promotion clearly suggests that the University of Wisconsin was acting in legal self-defense in an effort to demonstrate its commitment to gender equity on the faculty during a time when personnel practices in higher education were coming under increased public scrutiny and legal attack.[15]

Rudin was at first reluctant to accept the promotion:

Actually, Walter talked me into taking it; he said that it was insurance he couldn't buy for me. If something should happen to him and I had those four children, then I would be in trouble because I didn't have a position anywhere. And it was at a time, 1970, when jobs were actually very scarce. So I decided to accept if they would let me be part-time; I didn't want a full-time job. (Rudin: 19)

A few years later, she finally agreed to take the professorship full-time.

Mary Ellen Rudin's career is extraordinary in several respects. The security of Walter's employment and his support of her mathematical interests gave her the freedom she needed to continue her research without having to work full-time. Moreover, she has never based her sense of self-worth as a person or as a mathematician on external factors such as titles and tenure. At the same time, she is quick to acknowledge that the security and status of a professional position have been crucially important to many other women mathematicians of her own generation and later. The experience of Jane Cronin Scanlon provides an illustrative example.

Cronin Scanlon, like many of her fellow graduate students at Michigan, had dreams of going to Princeton on a postdoc, as her teacher Warren Ambrose had done. She was unsure as to whether postdocs were open to women, however, until (by a happy accident) she discovered that Dorothy Maharam Stone had held one:

I went to the library one day, and there was a new journal. And I looked at the table of contents, and there was an article by Dorothy Maharam. So I looked it up: you know, it was something written by a woman. And there was a footnote at the bottom of the title page that said that this work had been done while she was supported by a postdoctoral fellowship. And I thought, "Well, *she's* a woman; *she* got a postdoctoral fellowship. *I'm* a woman; *I'll* try it." And I did get one. (Cronin Scanlon: 16)

Cronin Scanlon's postdoc, sponsored by the Office of Naval Research, supported her for one year in the mathematics department at Princeton.

Early in her career, Cronin Scanlon was not particularly interested in teaching; she believes her parents had something to do with this. "They didn't quite approve of teachers," she says. "I think they may have had in mind, you know, George Bernard Shaw's remark, 'Those who can, do; those who can't, teach'" (Cronin Scanlon: 21). Because she loved the research she was doing, because she wanted to be a "doer" rather than a "teacher," Cronin Scanlon worked in a succession of research positions after her Princeton postdoc, first at Michigan and then at Courant. Finally, in 1951, she found more permanent employment at the Air Force Cambridge Research Center outside Boston, where she remained on the staff until 1954. During the five years that followed the completion of her Ph.D., Cronin Scanlon developed the *habit* of research and publication—a habit that proved difficult to break when, in later years, her life circumstances became more complicated.[16]

In 1953, she married Joseph Scanlon, a physicist and colleague at the Research Center; she gave birth to their first child the following year. Marriage and childbirth imbued her with an acute sense of responsibility. "After I was married," she says, "I realized that one had to think of other things than what one was doing mathematically!" (Cronin Scanlon: 23). In particular, she had to find a way to reconcile her desire to stay home and care for the baby with the family's very real need for her income. The dilemma led her away from her job at the Air Force Cambridge Research Center and back into college teaching. During the academic year 1954–55, Cronin Scanlon took two part-time teaching positions—one at Wheaton College, the other at Stonehill College—that were an easy commute from her home south of Boston.

The year 1956 was a particularly tumultuous one for the Scanlon family. Their second child was born in April; during this same period, they made a succession of moves—first to south-central Massachusetts, then to New York City, and finally to suburban New Jersey—as Joseph Scanlon made a succession of job changes. The move to the New York metropolitan area came too late for Cronin Scanlon to find a teaching job for the fall of 1956, but in January of 1957 she was hired as an assistant professor at the Polytechnic Institute of Brooklyn.[17]

During her first term of full-time teaching at Brooklyn Polytechnic, Cronin Scanlon taught a twelve-hour load. Between teaching and child care responsibilities, she seemed to have little time for research: "It seemed to me that I had certain responsibilities. If I had any time left over after the discharging of those responsibilities, then I'd do mathematics" (Cronin Scanlon: 24). Nevertheless, it was at about this time that Cronin Scanlon wrote a paper on topological methods in ordinary differential equations, a copy of which she sent to Solomon Lefschetz at Princeton. Lefschetz, with whom she had been acquainted during her Princeton postdoc, offered her some research grant support. The grant money allowed Cronin Scanlon to teach a reduced course load at Brooklyn Polytechnic and to spend some uninterrupted time working on her mathematics at Princeton. She describes her visits to the Princeton campus as having afforded her "the first time I'd been able to sit down for four hours at a stretch" since the birth of her first child; "it was tremendous, it really helped me infinitely" (Cronin Scanlon: 28).

Cronin Scanlon insists that she never planned to have a *career* in mathematics. She says, perhaps only somewhat in jest, that she is incapable of thinking more than three months into the future and that she rarely makes long-range plans. Certainly the daily rhythms and the unpredictability of life with small children make such planning very difficult, if not impossible. All of these considerations notwithstanding, by 1958, Jane Cronin Scanlon was in the midst of a very productive mathematical career. In that year, she became—to the best of her knowledge—the first woman to be awarded tenure at Brooklyn Polytech, and she gave birth to her third child.

In the years that followed, Cronin Scanlon set out to obtain grant support on her own, with great success; her research was funded by both ONR and NSF. During the academic year, she relied on the services of a babysitter who stayed with the family; in the summer, she used the babysitter part-time. Almost all of her mathematical research and writing were done at home. "I retired to the bedroom," she recalls, "and I only came out if there was an emergency" (Cronin Scanlon: 34). Her work was done efficiently during scheduled work sessions and always during daytime hours. If it is true that Cronin Scanlon never planned her career, it is nevertheless the case that she planned her days, weeks, and months so that her family, her teaching, and her mathematics would all reliably receive the attention they deserved.

By 1965, Jane Cronin Scanlon had four children; she was a tenured full professor at Brooklyn Polytechnic; and she had published a critically acclaimed book, *Fixed Points and Topological Degree in Nonlinear Analysis,* which attracted the attention of both physicists and mathematicians, particularly at nearby Rutgers University. In the fall of 1965, Cronin Scanlon accepted a tenured full professorship at Rutgers, where she remained on the faculty until her retirement in 1991.

Kenneth Wolfson was the chair of the mathematics department at Rutgers who made the offer to Cronin Scanlon in 1965. At about the same time that Cronin Scanlon was hired, the NSF awarded a Center of Excellence Grant to Rutgers for the improvement of its research and graduate programs in mathematics and the sciences. The grant provided Wolfson with $1 million with which to make new faculty appointments during the years 1965 to 1968. Wolfson was looking to recruit experienced mathematicians with established research reputations at a time when such people were greatly in demand. While other universities were still reluctant to hire women into research-oriented positions, Wolfson actively recruited them, and Cronin Scanlon was among his first female hires. When, in the late sixties and early seventies, legal pressures mounted against those universities who had not previously hired women, the mathematics department at Rutgers was held up as a model example of progressive hiring practices and gender equity.[18]

Unlike Mary Ellen Rudin, Jane Cronin Scanlon seems to have kept the various parts of her life tightly compartmentalized. While she did much of her mathematical research at home, she largely insulated her husband and children from her mathematical work. On campus, she said little about her family life. "When I was at Brooklyn Polytech I realized it would be a good idea not to mention the fact that I had two small children," she recalls. "I think that was a very sensible conclusion." Moreover, despite the many political gains that women have made since the late sixties, she holds to her belief that, because motherhood tends to be viewed as a liability in the workplace, it is a bad idea for a woman faculty member to discuss her family life at work. In that regard, she says, "I don't think anything has changed" (Cronin Scanlon: 39).

This strict separation of the personal and professional points to another key difference between Mary Ellen Rudin and Jane Cronin Scanlon. Walter and Mary Ellen Rudin worked in research areas that were very far removed from one another (complex analysis and set-theoretic topology, respectively), and each established a reputation in research independently from the other.[19] Yet Mary Ellen Rudin's colleagues were always well aware that she was Walter's wife, and thus they saw her in a female role with which they could feel comfortable. This role afforded her considerable freedom to pursue her mathematical research—but it also delayed her promotion to a regular faculty position.

By contrast, by maintaining a strict wall of separation between her work and family lives, Jane Cronin Scanlon presented herself as a professional to be reckoned with on her own terms. The effort to make mathematics a central part of the fabric of her life was clearly not without its costs. Cronin Scanlon was occasionally chastised by women in her church who did not work outside the home for "leaving her children." Although her husband was supportive of her professional activities and pleased by her economic contribution to the family, Cronin Scanlon says, "We were still living at a time when for the woman to work was a threat to a man's ego" (Cronin Scanlon: 30). While the marriages of Rudin and Morawetz withstood the extraordinary demands of a large family and dual careers, Cronin Scanlon's marriage ended in divorce in 1979.

One Step Forward, Two Steps Back

For many of the interviewees, the struggle to balance work, marriage, and children brought on a succession of small triumphs and often daunting setbacks, as achievements in one area of life were often followed by crises in another. Very few of these women enjoyed the luxury of continuous career development as outlined in the myth. Indeed, for most of them, a sense of personal and professional equilibrium did not come until fairly late in life, as the experiences of Tilla Weinstein serve to illustrate.

Weinstein married for the first time in 1953 while still in college. I have already related the story of her early pregnancy and of the efforts that she and her first husband made to share household and child-care responsibilities while both were in graduate school at NYU in the fifties. Her initial job search, on completion of the Ph.D., stands in striking contrast to those of the other women interviewed. Weinstein's adviser, Lipman Bers, clearly regarded her as a promising research mathematician, and through his own old-boy network made every effort to place her in a research-oriented department of mathematics. Moreover, Bers was among the first to recognize what has come to be known as the two-body problem: the problem of finding satisfactory academic employment for *both* spouses. Weinstein recalls that "since [my husband's] was the harder job to get—positions in mathematics were a dime a dozen then!—Lipman Bers's *dream* was that *he* would get a job in English for my husband! He would have *loved* that!" (Weinstein: 20).

In 1958, her dissertation essentially complete, Weinstein was offered an instructorship at the University of California at Berkeley, secured through Bers's network of personal connections. The position carried responsibilities in both teaching and research; it included financial support for her attendance at meetings and conferences, along with a moving allowance. In short, it was the sort of contract that would be offered to any *man* embarking on a research career in mathematics. Unfortunately, Weinstein's husband was unable to find a job in the Berkeley area, nor was he quite as far along on his dissertation as she was; so she turned down the Berkeley offer, and they remained in New York for another year.

Bers was every bit as devoted to landing a good job for Tilla Weinstein the following year, but this time he urged her to follow the more traditional strategy of letting her husband get his job first and then trying to find a job for herself in the same area. When her husband found a job at San Fernando State College (which soon afterward became Cal State Northridge), Bers helped Weinstein to secure an instructorship at UCLA.

Initially, the environment at UCLA was less than ideal. The instructorship, which was nominally to last only one year, carried an eleven-hour teaching load. While her colleagues seemed eager to welcome her into the *social* life of the department, it was all but impossible to find colleagues who wanted to interact with her *mathematically*. While her instructorship was renewed for another year, her mathematical isolation continued; in response, she recalls, "I just started my habit of working on my own" (Weinstein: 20).

Early in the fall of her second year at UCLA, her colleagues met to discuss Weinstein's future in the department. "Angus Taylor—who was chairman, and the *epitome* of a fair-minded man," called Weinstein in to discuss what had gone on at the meeting:

I remember his eyes popping open as he said, "You seem to have made a lot of *friends* here. . . . But I have to be honest; you should know everything that was said. One of our colleagues got up and said, 'Gee, you know, it's *awfully* nice having her around *now*. She's really very pleasant. But twenty years from now, she may just be a grouchy old lady!'" To which I said—one of the few times in my life that I've ever said *exactly* what I would have wished I would have said— I said, "Well, I can't very well turn into a grouchy old *man*." And his reaction was one of complete blankness. It was as if I hadn't said anything; it did not compute. He did not hear in what he had said anything except the report of a statement by a colleague. He didn't see why I said what I said because I certainly said it in a soft tone and not in a challenging tone. (Weinstein: 22–23)

Utterly unfazed by Weinstein's response, Taylor continued. He asked what her future research plans might be, and, as she recalls it, he asked whether she planned to have more children. "Well," she replied, "I might. But you can look at my record and see whether having one child in any way caused me to stop in my career. And if I have a second child, I'm in a better position now to finance the care of the child and continue" (Weinstein: 23).

Shortly after this meeting, Weinstein sensed a sea change in the department's attitude toward her:

They very quickly decided to make me an assistant professor . . . on a regular tenure-track line. . . . And from that moment on, I never sensed the slightest bit of awkwardness in the department. Now, for all I know, one or two members of the department were uncomfortable with it, but so what? It just wasn't there any more as an issue. (Weinstein: 23)

From then on, Weinstein was treated like a regular member of the department. She was given a reduced teaching load and time for her research; NSF support freed her from the need to teach summer school. By contrast to her earlier experience at NYU, she was allowed to continue teaching right on through her second pregnancy, which culminated in the birth of her second son in July of 1962. When, in 1964–65, she took a year's leave of absence to do research at NYU, her *husband* was the one to follow, taking a visiting position at Queens College. On her return, in 1966, Tilla Weinstein became the first woman to be awarded tenure in the mathematics department at UCLA.

But at the very moment of professional triumph, she found her marriage "very suddenly and unexpectedly coming to an end" (Weinstein: 26). The divorce marked the beginning of a period of great personal and professional upheaval. Her sons continued to live with her—"it would have been rare for it to be otherwise in those days," she says—but she recalls feeling "unhappy" and "imbalanced . . . in the sense of being not completely yourself and still a bit off-stride. I think I remained really affected by the end of that marriage for many, many years" (Weinstein: 27).

Her 1968 marriage to John Milnor (Ph.D., Princeton, 1954)—then temporarily on the UCLA mathematics faculty after going through a difficult divorce of his own—added another layer of complication. To be closer to his children, Milnor needed to move back to the East Coast. To accompany him, Weinstein eventually resigned her hard-won position at UCLA, where she had worked so hard to establish herself and had been a trailblazer for women in the process.

Her first few years in the East were unsettled. Weinstein held a succession of temporary positions, while both sets of children, disrupted by the dramatic changes in their lives, required a great deal of time and attention. When Milnor accepted a permanent position at the Institute for Advanced Study, Weinstein was determined to secure a permanent position of her own in the Princeton area. In 1970, she was hired as chair of the mathematics department at Douglass College, the women's coordinate college of Rutgers University. "It seemed to me," Weinstein says, "that within an easy commute of Princeton, Rutgers was the only school of the appropriate quality, both in terms of what I could offer them and what they could offer me." While she was clearly apprehensive about taking on an administrative post, she adds, "It was not something I was going to agree to do and then treat lightly" (Weinstein: 29).

This was Tilla Weinstein's first administrative position, and as such it was a dramatic departure from her previous professional identity as a researcher and teacher. Although she had much less time for research at Douglass than she had had at UCLA, her research accomplishments were regarded with respect. She was appointed to the graduate faculty at Rutgers, which meant that she could teach graduate courses and supervise dissertations there. She credits Kenneth Wolfson, who was still the chair of the Rutgers department, with ensuring that she be made welcome. "Wolfson," she recalls, "wanted very much to have influence in the Douglass math department" and generally wanted to ensure that mathematics at Rutgers was of the highest quality possible (Weinstein: 30).

Weinstein spent the years 1970 to 1973 as department chair at Douglass and was responsible for hiring many outstanding mathematicians, male and female, who ultimately joined the department at Rutgers when the faculties merged in 1980–81.[20] Weinstein was, by all accounts, a talented department chair, but by the mid-1970s she was more than happy to return to research and teaching. It was at this critical juncture that her marriage to Milnor began to collapse. During the academic year 1975–76, they divorced. Reflecting on that marriage, Weinstein says simply, "Jack is very sweet. The two of us just aren't people who ever should have been married to each other" (Weinstein: 28).

In geographical and professional terms, Weinstein's second divorce was much less disruptive than the first. She remained on the faculty at Douglass, and she continued to live in New Jersey with her younger son. But mathematically and personally, she felt herself to be at a crossroads: "At that point, I felt I had to remake myself in many ways" (Weinstein: 38). In the years that followed, she reestablished her relationship to mathematical research, as if from scratch. She remained unmarried until 1991, in itself a remarkable change for a woman who had never *not* been married. And, in the 1980s, while in her forties and fifties—and in contradistinction to the myth of the mathematical life course—she came into full flowering as a research mathematician.[21]

Professional Success, Marital Failure

For Rebekka Struik and Vera Pless, as for Tilla Weinstein, divorce followed closely on the heels of professional success. Whereas Weinstein's first marriage was remarkably egalitarian for its time, both Struik and Pless spent many years following their husbands' careers, bearing and raising children, and struggling to establish satisfying professional lives of their own. Their marriages came to an end just as they were attaining the professional recognition they had worked for years to achieve.

Rebekka Struik and Hans Freistadt married in graduate school, and from the outset of their marriage, Struik followed Freistadt through a succession of often unpredictable career moves. Trained as a physicist, Freistadt's prospects in physics were sharply curtailed when he was blacklisted for his leftist politics during the McCarthy era. By 1957, he had decided to seek his fortunes elsewhere and enrolled as a student at the University of British Columbia Faculty of Medicine. Struik recalls, "I married a physicist and divorced a physician" (Struik: 14).

While offering support and encouragement to her husband as he made a major career change, Struik had vitally important ambitions of her own. Her mother had been unable to successfully combine mathematics and motherhood. Haunted by the memory of her mother's frustration, Struik was bound and determined to raise a family *and* secure a tenure-track position in a university mathematics department. At British Columbia, however, the mathematics department refused to take her seriously, regarding her as merely "a student wife" seeking temporary employment (Struik: 18). When the family moved to Colorado in 1961 for Freistadt's internship and residency, Struik readily found college teaching positions, but a tenure-track job remained elusive.

Struik recalls her husband's reassurances that, once his residency was over, he would move with her "anywhere . . . in the country" so that she could take "a really *good* job" (Struik: 19). Yet they remained in Colorado, and it was there, in 1964, that Struik landed her first tenure-track job, at the University of Colorado at Colorado Springs. By this time, however, the family—which now included three daughters—had moved to Pueblo, where Freistadt had taken a job as county physician. For Struik, working toward tenure meant commuting between Colorado Springs and Pueblo, "[coming] home Wednesdays and weekends" (Struik: 20).

When, in 1968, Struik had at last earned tenure, the family moved yet again—this time, to Denver. Fortunately, Struik was able to transfer her tenured position to the University of Colorado's campus there. But her years of commuting had taken a toll on the marriage, and in 1969, she and Freistadt were divorced. The following year, Struik transferred her position back to the University of Colorado at Boulder, where she has remained ever since. She has not remarried.

The experience of Vera Pless provides a variation on a similar theme. She married in 1952, just as she was completing a master's degree in mathematics at the University of Chicago. Unlike Struik, she at first had no intention of pursuing a profession of her own. Her husband, Irwin Pless, was working on a doctorate in physics, and when she married him, her plan was simply "to wait for him to get his Ph.D." (Pless: 7). What Pless had not anticipated, however, was the motivating power of boredom. Hers was a vital, active, creative mind, a mind that could not bear to remain idle. It was mental restlessness that drove her to return to graduate school, completing a Ph.D. of her own at Northwestern while she "waited" for her husband to finish up at Chicago.

Irwin Pless completed his doctorate first and immediately took a position in the physics department at MIT. Vera Pless moved with him to the Boston area, where she completed her Ph.D. *in absentia* and gave birth to her first child two weeks after her dissertation defense. During her early years in Boston, Pless got a part-time teaching job at Boston University. She didn't do it for the money; she didn't see herself as pursuing a career; and she certainly didn't do it for the prestige. "I sought it because I really thought that my mind was going," Pless recalls with a laugh, "being home with the kids" all day (Pless: 11).

It only gradually dawned on her that creative intellectual work in mathematics was something she really couldn't live without. After four years of child care and part-time teaching, unable to find a job in academia, Pless took a job at the Air Force Cambridge Research Laboratory (where Jane Cronin Scanlon had worked nearly a decade before). During her ten-year affiliation with the lab, she raised three children, with the help of babysitters and nannies; continued as a supportive spouse to her husband; and became one of the world's leading experts in the brand-new field of coding theory, the field in which she ultimately made her mark as a mathematician. Pless's pioneering work in error-correcting codes found a wide interdisciplinary audience—in electrical engineering, computer science, and mathematics; in academia, government, and industry. But the passage of the Mansfield Amendment, whereby the armed services were no longer permitted to engage in the support of pure research, brought Vera Pless's association with the air force lab to an abrupt end in 1972.[22]

"It was a very traumatic experience for me," Pless recalls. Although she was offered the opportunity to remain at the lab, "doing something very applied in the armament section," she could not bear to leave the work she had been doing "and just left" (Pless: 15). It was a particularly inauspicious time for her to begin an academic job search. In the early seventies, college and university faculty positions were extremely scarce. Although the passage of Title IX meant that faculty appointments were increasingly available to women, progress was slow and incremental. Moreover, Pless had few personal contacts in academia, having been disengaged for fifteen years from the mainstream of the mathematical community. To further complicate her reentry into academic life, Pless restricted her job search to the Boston area.

Ultimately, Pless won an appointment as a research associate in the electrical engineering department at MIT. The project to which she was assigned—the development team for MACSYMA, one of the first symbolic algebra programs—was not terrifically interesting to her. But it afforded her the opportunity to do some teaching and sufficient free time to continue her coding theory research; she also applied for and was awarded her first NSF grant. By 1975, she was determined to land a professorship of her own, having come to see the achievement of this goal as "vital to my well-being" (Pless: 18). This time she applied for jobs not merely in the Boston area but all over the country. She felt that it was time for her husband, who had enjoyed nearly twenty years at MIT, to consider the possibility of relocating for the sake of *her* career.

Vera Pless's 1975 academic job search landed her a tenured full professorship in mathematics and computer science at the University of Illinois at Chicago. "There were other options," Pless recalls, "but this looked like the best to me. And somehow going back to Chicago appealed to me at that point" (Pless: 21). With the understanding that her husband would eventually find a job nearby and join her, Pless left the family in Boston and came to Chicago to accept her first "real" academic job.

In a marriage where the husband's career has always taken top priority, it is often unusually difficult to shift the balance later on—to give precedence to the wife's career and to allow her to take a turn. As it happened,

Vera Pless's husband remained on the faculty at MIT and did not move to Chicago. For five years, the Pless family was based in Boston as Vera commuted back and forth to and from her job in Chicago. Commuting didn't "kill" the marriage, Pless recalls, "but it didn't help" (Pless: 22). Their marriage ended in divorce in 1980.

The Roads Not Taken

In this chapter I have discussed a variety of ways in which the women mathematics Ph.D.'s of the 1940s and 1950s struggled to make mathematics an integral part of their lives. Indeed, most of the women interviewed maintained an ongoing professional involvement in mathematics while at the same time carrying out a host of other roles and responsibilities. Several interviewees, however, found that their career paths diverged—sometimes dramatically—from what they had originally envisioned for themselves. Still others left mathematics and mathematical activity for unusually long periods of time to concentrate on personal and family responsibilities, only to return to mathematics much later in life. And one interviewee—Jean Walton—made the decision to leave mathematics entirely at a fairly early stage in her career and never returned.

When Margaret Willerding was still a graduate student at St. Louis University, she became interested in number theory and demonstrated genuine talent for mathematical research. She recalls her dissertation supervisor, Arnold Ross, with fondness as "not the best teacher but a delightful man" (Willerding: 5). Ross left St. Louis to chair the mathematics department at Notre Dame University in 1946, before Willerding's dissertation was complete; so during her final year of graduate study, she regularly commuted back and forth to South Bend to work with Ross. At Notre Dame, she spent all of her time with Ross, his wife, and Gene Guth—a theoretical physicist from Hungary and a close friend of Ross—to whom she was briefly engaged to be married.[23]

"I was trained to be a research mathematician," she says. But Ross and Guth were the only research mathematicians she knew, and the narrow focus of their lives, which conformed rather neatly to the myth of the mathematical life course, made her question the desirability of becoming a research mathematician herself. "All they did was eat, drink, and sleep mathematics, and I said, 'There's more to life than this,'" Willerding recalls. "They thought it would be very chummy if I married Guth—he was about twenty-five years older than I—and [came] up to Notre Dame. . . . We would just be a good little group there. But I didn't see it that way" (Willerding: 5–6).

Her misgivings about having a research career notwithstanding, Willerding did not immediately rule out the possibility. On completing the Ph.D. in 1947, she wrote up her dissertation for publication in the *Bulletin of the American Mathematical Society,* and she was hired as an assistant professor by Washington University in St. Louis, which at that time was working to build up its research faculty.

But her experiences during her brief stay at Washington University turned her decisively away from the world of pure mathematics. She felt, in particular, that she was treated as a second-class citizen because of her gender from the moment she arrived. To begin with, the department head informed her on her arrival that she would not be promoted "as fast as a man, even if I did as much or more work than they did" (Willerding: 7). On another occasion, when a regional meeting of the AMS was held at Washington U., Willerding recalls: "One of the faculty wives called me up and wanted me to pour at one of the teas they were having. And I said, 'I don't intend to pour at one of the teas you're having. I'm one of the *faculty*'" (Willerding: 11). After just one semester at Washington University, Willerding resigned.

Having been previously employed in the St. Louis city school system, Willerding requested a reactivation of her employment there in early 1948. Because she was one of the few teachers in the system with a Ph.D., they honored her request to be assigned to the faculty of her undergraduate alma mater, Harris Teachers College. And with these simple steps, Willerding left the world of research mathematics, never to return. For the next thirty years—first at Harris Teachers College and later at San Diego State University—she devoted herself to the study of mathematics education and to the preparation of elementary and secondary school mathematics teachers. She made a conscious choice to remain unmarried, enjoying the companionship of close friends and the freedom to write and to travel.

Nearly fifty years later, however, Willerding still expresses some ambivalence about her decision to leave mathematical research behind. On the one hand, she asserts plainly that she *preferred* to teach in "a state university where they didn't expect me to do research in mathematics and do *nothing* but research" (Willerding: 5). On the other hand, she speculates with a laugh, "If I could have gone to Bryn Mawr and studied with Emmy Noether, I might have been quite a famous mathematician now" (Willerding: 12). This cryptic remark suggests that, had Willerding had even *one* role model to show her how research mathematics can be integrated into a *woman's* life, she might have chosen differently. She clearly had the *ability* to succeed as a research mathematician. Moreover, she also had a tremendous capacity for creative work, as her later output makes clear: she is the author of over two dozen textbooks on high school and college mathematics, many of which have had multiple printings and have been translated into several languages. While she has done very fine work in mathematics education, her words reveal her deep disappointment at the mathematical research community's inhospitality to women.

Although Margaret Willerding never had the mathematical research career for which her doctoral studies prepared her, she nevertheless had a long and productive career as a teacher and writer. Several of the other women interviewed, however, never had careers commensurate with their training and abilities. Janie Bell stands out as one such woman who seems remarkably happy with the way her life turned out.

Clarence Bell and Janie Bell both completed their Ph.D.'s at Illinois in 1943. Shortly thereafter, Clarence took a position at Battelle Memorial Institute in Columbus, Ohio, in lieu of being drafted into the army. At the war's end, Clarence and Janie briefly discussed the possibility of trying to obtain dual teaching positions on the mathematics faculty of a liberal arts college. This kind of job search, while commonplace now, was unheard of then. For whatever reason, the Bells quickly abandoned the idea and Clarence decided to stay at Battelle. From the mid-1940s through the mid-1960s, Janie raised three children while teaching on an as-needed basis at Ohio State University and at a nearby high school as well. After Clarence Bell's death in the late 1960s, she briefly held teaching positions at two Presbyterian-affiliated liberal arts colleges in North Carolina, Montreat-Anderson and Warren Wilson; in recent years, she has served as both classroom teacher and individual tutor for elementary and secondary mathematics students. Proud of her children and of her involvement with teaching at all levels, Janie Bell seems generally very content with the extent of her involvement with mathematics.

By contrast, Anne Calloway, whose career as an academic mathematician effectively ended with her marriage in 1952, has struggled for years with the consequences of her departure from academia. Her attempts to work in isolation on research problems proved futile, and over time she withdrew entirely from teaching, feeling that she wasn't really cut out for the job. As her children grew older she contemplated the possibility of nonacademic employment in a field that would make use of her mathematical talent and technical skills. In the late 1970s and early 1980s, she returned briefly to graduate study, in computer science this time, and worked for several years as a programmer for the state department of transportation in Michigan. While she maintains an active interest in mathematics and computer science, Calloway seems disappointed that she never really made her mark in either field.

By contrast to Bell and Calloway, Mary Dean Clement *did* sustain a mathematical career that lasted for well over twenty years post-Ph.D. Despite the continuity of her employment, however, Clement never achieved a high degree of satisfaction in any of the positions that she held. During her eight postdoctoral years on the faculties of Wells College and the University of Miami, Clement was repeatedly frustrated in her attempts to do research by the burden of heavy teaching loads and disappointed by the sense that her contributions as a teacher went unappreciated by her colleagues and supervisors. No longer able to tolerate the humid Florida climate, Clement resigned her position at Miami in 1949 and returned to Chicago with the goal of making her way back into mathematical research. During the years that followed, she held temporary teaching positions at Chicago and Northwestern and did, indeed, publish her first mathematical research paper, in the field of geometry, which she had studied as a graduate student under Lane. But she was unable to sustain her research momentum, and—weary from her repeated job searches—she opted for a job at the Institute for Air Weapons Research at the University of Chicago. She continued to work there until health concerns brought on her early retirement in the mid-1960s.

The experiences of Joyce Williams stand in striking contrast to the stories of women who were unable to fulfill their early promise and clearly shows that women are fully capable of returning to productive mathematical lives after a lengthy hiatus. It will be remembered that, in the fall of 1950, Williams and her husband entered graduate school at Illinois, where he had an assistantship and she did not. At the end of that first academic year, their fortunes reversed: she won a fellowship, while his was not renewed. While her husband moved to Buffalo to take a job, Williams stayed on until 1954, completing her Ph.D. with Waldemar Trjitzinsky. By the time she was ready to graduate, her husband had landed a job in Massachusetts with Raytheon, and Williams decided to look for teaching jobs in the Northeast. She remembered receiving two replies, one of them from Tufts University, which simply stated that "they did not hire women" (Williams: 18).[24] The other reply, from Smith College, was more promising.

"Smith was interested," Williams recalls, "but by the time I realized I was pregnant, *I* wasn't interested" (Williams: 19). So she chose to forego the possibility of a faculty position at Smith and returned to Massachusetts to start a family. Between 1954 and 1959, she had four children, one of whom was diagnosed with leukemia and died at the age of nine; in the late 1960s, she gave birth to a fifth child. Save for one year in which she taught at nearby Lowell Technological Institute, Joyce Williams devoted her time and energy in the fifties and sixties to the responsibilities, joys, and sorrows of family life.

In view of the intensity and tragedy of these early years, what she went on to do in the seventies and eighties seems all the more remarkable. In 1973 she was hired to teach mathematics full-time at Lowell Tech (subsequently renamed the University of Lowell and, most recently, the University of Massachusetts Lowell). In 1977 she was promoted to assistant professor and given a tenure-track position. Tenure would be awarded, she was told, if she produced one substantial piece of published research. Nearly twenty-five years after the completion of her Ph.D., Williams dusted off her dissertation, reacquainted herself with its contents, revised it, and wrote up a paper on singular integrals with complex-valued kernels, which was published in the spring of 1979. Inspired by the success of this effort, she actually continued to do research on integral equations for a short while longer. The subsequent papers she wrote on the subject were, unfortunately, never published. Her return to research, short-lived though it was, was nevertheless sufficient to merit tenure at Lowell, where Williams remained on the faculty as an associate professor until her retirement in 1996.

While Joyce Williams took a long, but ultimately temporary, hiatus from mathematics, Jean Walton's hiatus from mathematics was deliberate and permanent. Unlike most of the other interviewees, Walton's relationship to mathematics can best be described as having been ambivalent from the very start. When she began her doctoral work at Penn in the mid-1940s, she was uncertain as to whether she would pursue a career in administration or in teaching. But Walton's experiences at Penn—her difficult relationship with Hans Rademacher and her intense isolation from students, faculty, and the wider mathematical community—served mainly to reinforce her longstanding belief that a mathematical career would be lonely and arduous.

When her doctoral dissertation was at last complete, Jean Walton's ensuing job search was conducted in a rather peculiar fashion. Although she was seeking a faculty position in mathematics, she did not ask Rademacher to use his connections to help her find a job. Rather, she asked the president of Swarthmore College—a former philosophy professor who was familiar with her work as both student and assistant dean—to write to the presidents of other small colleges on her behalf. It is therefore not surprising, then, that when a job offer came from Pomona College, it was for the position of dean of women—rather than, say, an assistant professorship in mathematics. While she continued to teach mathematics part-time at Pomona for another ten years, Jean Walton grew increasingly disconnected from mathematics. While she *did* take the crucial step of promptly publishing her doctoral dissertation, this was, perhaps, both her first and last act as a professional mathematician.

7

Teaching, Research, and the Question of Identity

Becoming a Mathematician

What does it mean to be a mathematician? And what does it take to come to identify oneself as such? For Paul R. Halmos (Ph.D., Illinois, 1938), whose mathematical career came into full flower during the period he calls "the fabulous fifties," the answer is simple. "[Y]ou have to be born right," he says, "you must continually strive to become perfect, you must love mathematics more than anything else, you must work at it hard and without stop, and you must never give up" (Halmos 1985: 400). Halmos's view—which, he confidently asserts, would be shared by many if not most of the outstanding creative mathematicians of the twentieth century—expresses the central tenet of the myth of the mathematical life course: the process of becoming a mathematician is one of single-minded, lifelong dedication to the ideal of research excellence.

Not surprisingly, Halmos's views stand in sharp contrast to the experiences, thoughts, and feelings of the women who received mathematics Ph.D.'s from American universities in the 1940s and 1950s. Like Halmos, nearly every one of the interviewees has been involved in significant mathematical activity for the better part of her adult life. Unlike Halmos, however, these women have had neither the time nor the inclination to focus on mathematics to the exclusion of other social, personal, and intellectual concerns. Moreover, for each of these women, becoming a mathematician meant creating a *modus vivendi* that minimized the conflict between the two cultures of postwar American mathematics: the lower-status, female-identified culture of teaching, and the higher-status, masculine culture of research. For some, this meant identifying exclusively or predominantly with either the teaching or the research role; for others, it meant striking a compromise between the two. And for a small minority, this meant moving beyond the narrow categories of teaching and research and into new areas of creative activity.

In this chapter, I explore the variety of professional roles taken on by the women mathematics Ph.D.'s of the forties and fifties. In particular, I examine how these women responded to the tension between teaching and research and how, in the process, they developed a sense of themselves as mathematicians.

Research Women

As discussed in chapter 2, nine of the interviewees clearly identify themselves primarily with the research role. All nine have been involved in teaching, and some have held important administrative leadership posts; but mathematical research is the activity they talk about the most in their interviews, and research is their top professional priority. Thus they have enjoyed a high degree of consonance between their own goals and the values of the postwar mathematical community. At the same time, the unfolding of their lives and careers has diverged sharply from the timetables set out in the myth of the mathematical life course. Moreover, they hold subtly different views regarding their sense of identity as mathematicians.

Within this group of research women, attitudes toward teaching vary widely. For Cathleen Morawetz and Jane Cronin Scanlon, the disjunction between research and teaching is quite sharp. Morawetz states plainly that in the early years of her career, "I did not want to teach. I thought that if I took on teaching, I would not be able to also do research." She seems pleased that, when NYU finally invited her to join the mathematics faculty in the late 1950s, the offer came from the Courant Institute, "where they taught only graduate courses," rather than the undergraduate college. She admits, "Frankly, I prefer administration to teaching" (Morawetz: 18). Despite similar early misgivings, Jane Cronin Scanlon turned to college teaching for pragmatic reasons: it helped to pay the bills. Although she eventually came to like teaching, she is inclined to view it—along with other professorial responsibilities—as "just a job," as distinguished from the activity of *doing mathematics,* which she continues into retirement (Cronin Scanlon: 43).

Mary Ellen Rudin has always held that teaching is "a natural part of being a mathematician," but she, too, sees teaching and research as essentially distinct activities. Unlike her mentor, R.L. Moore, for whom every mathematics class was an apprenticeship in research, Rudin did not believe it was her place to "sell mathematics in the classroom" (Rudin: 6). By contrast, Domina Spencer regards teaching, both undergraduate and graduate, as a means of communicating, clarifying, and refining her research ideas. "I learn so much from trying to explain my research to my students," she says, "that it would be a great disadvantage not to have them" (Spencer: 41).

All of the research women are or have been married, and all have had children. Although many of them continued to do research while their children were young, even the most dedicated found it difficult to do mathematics during periods of intense family responsibility. Thus it is not surprising that these women generally did their best work at middle age and beyond. Even Mary Ellen Rudin, blessed with extraordinary self-confidence and an astonishingly high tolerance for distraction, acknowledges that her mathematics improved with age. "My mathematics became best when I was about sixty," she says and adds that this is an experience common to many women of her generation (Rudin: 26).

Not surprisingly, the women in this group generally agree that to call oneself a mathematician, one must be actively engaged in mathematical research. Patricia Eberlein draws a careful distinction between *mathematicians,* "a certain set of people that work on mathematical problems that think a certain way," and people who are *in mathematics,* whose primary activity is teaching. A mathematician is someone who is drawn, again and again, to the process of creating something new in mathematics, someone who enjoys "the crystal moment of great discovery" when a new idea first takes hold. Eberlein began to think of *herself* as a mathematician within a few years of completing the Ph.D., when she began to produce interesting research results on her own, without the guidance of an adviser (Eberlein: 14–15).

By contrast, Jean Rubin did not begin to think of herself as a mathematician until over *ten years* post-Ph.D. While still a graduate student at Stanford, she met and married the mathematician Herman Rubin (Ph.D., Chicago, 1948). On receiving the Ph.D. in 1955, she followed him as he took a succession of professorships, first at Oregon and later at Michigan State. Her primary responsibility during these years was the care and rearing of their two children, though she did hold temporary faculty positions, published four research papers, and coauthored a book on set theory with her husband. In 1967 (at the age of forty), Jean Rubin received her first offer of a regular faculty position when Purdue awarded tenured professorships to both Rubins. It was at about the same time, she says, that she first began to think of herself as a mathematician.

In the case of Tilla Weinstein, identification as a mathematician was an even longer time in coming. From her earliest days as a student at NYU, Weinstein felt herself to be an "outsider" in mathematics, "without a genuine instinct for the subject" (Weinstein: 19). Her classmates at NYU were what her then-husband referred to as *mathematical animals*. "They were creatures who by nature had an affinity for the subject, and it showed in the way they spoke and the way they looked at things, even if they had varying personalities" (Weinstein: 12).

But she was not one of them. She successfully completed a Ph.D. in mathematics; she earned tenure at a major research university; she maintained an active research program despite a tumultuous personal and professional life. But in Weinstein's view, simply doing research does not make one a mathematician:

> I didn't even think of myself as a mathematician—I didn't think of that word for myself—until quite recently. I'd say the first time that I felt that the mathematics I was doing was somehow, finally, flowing from inside of me, was right around 1980. . . . Before that, I *wanted* to be doing mathematics. I kept working; every now and then I really *found* something. But it was from plugging away and plugging away, and it was never seeming to come from inside of me. . . . I've always had a limited number of techniques that I could use. There wasn't a critical mass of mathematics in my subconscious. So I felt like someone who tried to do mathematics and every now and then proved something, to my great delight, and who loved teaching math and who enjoyed being around mathematicians. (Weinstein: 19)

For Weinstein, the process of becoming a mathematician is a psychological one, wherein mathematics becomes part of the subconscious. Unless one is naturally a "mathematical animal," this process takes time and comes to fruition much later for women than it does for men.

Jane Cronin Scanlon is unique among the research women in that she does *not* consciously identify herself as a mathematician. She states emphatically, "I don't think I ever identified myself as an *anything*." For Cronin Scanlon, mathematics is a body of knowledge, external to herself; doing mathematics is an activity, rather than an extension of the ego:

I'm pretty sure that I never think that I *am* a mathematician. Sometimes I *do* mathematics. Now, do other people think this way? I don't know. Certainly, I think some people—mercifully, not too many—use mathematics to the pleasure of their egos. You know, they need to show off. There are a few people who are really unpleasant—mercifully few.

When asked why she does mathematical research, Cronin Scanlon replies simply, "Because I like it" (Cronin Scanlon: 44–45).

For nearly all of the research women, "doing mathematics" is a lifelong passion, to be pursued whether or not one actually has a job in the field. Long before Dorothy Maharam Stone ever imagined becoming a mathematics professor, she was able to envision a life in which she worked in a department store by day and did mathematics at night. And although Maharam Stone and her husband formally retired from the University of Rochester in 1987, they share an office at Northeastern University in Boston, and both continue to work on mathematics. Similarly, Patricia Eberlein's fascination with problem solving continued into retirement, right up until her death in 1998 at the age of seventy-three. Jane Cronin Scanlon and Mary Ellen Rudin, now in their seventies, acknowledge that research does seem to take a little bit more time now than it used to, and Rudin foresees that there may come a day when she will decide to give it up entirely; but for the time being they show no sign of stopping. "Left to my own devices," Tilla Weinstein says, "I might never retire" (Weinstein: 42).

Teachers and the Pull of Research

In chapter 2, nine of the interviewees were readily identified as teachers of mathematics. These women did little research after the completion of the Ph.D., and teaching was by far their most important professional activity. Yet almost without exception, they are intensely aware of the fact that they have left research behind, and many of them have conflicted feelings about the comparative value of teaching and research. Do these women think of themselves as mathematicians, or, as Patricia Eberlein has suggested, are they merely women who have worked "in mathematics"? How do they feel about the place of mathematics in their lives?

Within this group of teachers, Margaret Marchand is the only one who actually made substantial research contributions in the early years of her career. In the early 1950s she published two papers in functional analysis, the area of her doctoral dissertation; from 1952 to 1956 she worked in cancer research and coauthored several papers in biostatistics. At least one of her results—a refinement of Student's T-test with important applications in oncology—was highly regarded by colleagues in the field, and she still remembers it with pride. But her return to college teaching in 1956 marked the end of her research career. She expresses no regrets about this transition. Indeed, she characterizes her teaching experience as having been "absolutely wonderful," at least until the point at which she felt that academic standards began to decline in the late 1960s and early 1970s (Marchand: 9).

It is interesting to note that Marchand has perhaps the strongest sense of self-identification as a mathematician of any of the women in the teaching group. When asked whether she considers herself to be a mathematician, Marchand replies, "I've been a mathematician all my life, and that's all that I can say." When pressed to say more about her identity as a mathematician, she affirms that her ability to "do mathematics"—to do research—has something to do with it, but quickly adds, "[mathematics] always came easy to me . . . so that's what I am" (Marchand: 22).

In addition to Marchand, both Anne Lewis Anderson and Evelyn Granville seem to be extremely content with their decision to make teaching their primary focus. Anderson spent just over twenty years on the faculty of the Woman's College of the University of North Carolina and headed the mathematics department during its transformation into the University of North Carolina at Greensboro in the early 1960s. The daughter of teachers and inspired by many teachers of her own, Anderson was a dedicated, enthusiastic, and much-beloved member of the faculty at Greensboro.

Thus it comes as something of a surprise when she adds, "I was kind of sorry I didn't try to pursue some more research than I did" (Anderson-1: 9). Anderson, who neither published her dissertation nor showed showed a particular inclination to go on in research, doubts that she ever would have found research to be as enjoyable as teaching. And yet her regret is tinged with shame: "I just think I didn't, perhaps, live up to my potential—you know, what I was trained for, the examples I had of my professors at Chicago and so on. I feel like maybe I let them down some" (Anderson-1: 10).

Evelyn Granville's attitude regarding the comparative merits of teaching and research is markedly different, shaped by her experiences of racial discrimination. From early childhood onward, she aspired to be a teacher; as a graduate student at Yale, she had her first introduction to research. Her experiences at NYU and Fisk gave her a taste of what a career devoted exclusively to either research or teaching might be like. But her employment opportunities as a African American woman in the 1950s were so severely constrained that she was unable to seriously pursue either option, turning instead to employment in government and industry. When, in the comparatively more liberal environment of the late 1960s, she once again sought employment in academia, she joined the faculty of California State University, Los Angeles. At Cal State L.A., her duties were primarily in undergraduate teaching, although she also branched out into mathematics education and involvement with the public schools. She seems happy to have been able to return at last to teaching, her first love. Indeed, so reluctant is she to give up teaching entirely that she has retired from not one but *three* different college faculty positions.[1]

For many of the other women who have devoted their careers to teaching, the tension between teaching and research evokes a much broader spectrum of emotions. For example, Joyce Williams says that when she was in graduate school she anticipated that she would go on to do a good deal more research. She seems somewhat regretful about what she calls "the big gap"—the twenty-five-year period between receipt of the Ph.D. and publication of her thesis. "I had thought originally I could keep up with [research] at home, [but] that just did not work out," she says. But her sense of disappointment is tempered by the realization that the choice she made to stay home and rear her children was made consciously and of her own free will (Williams: 15–16).

Complicated and ambivalent feelings about teaching and research often have their origins in graduate school, as we have seen in chapter 5. Alice Schafer, who felt that women were never given a fair chance to demonstrate their research potential at Chicago, is still very angry about it over fifty years later. In her career as a college mathematics teacher, she has invested an enormous amount of time and energy in encouraging women to go on to careers in research mathematics. She is determined to do whatever she can to ensure that future generations of women are given the kinds of opportunities that she feels were denied to her.

As graduate students at Wisconsin in the late forties and early fifties, Violet Larney and Augusta Schurrer could sense the growing chasm between the two cultures of teaching and research. Schurrer felt a natural affinity for teaching and feared that a research career might prevent her from enjoying a full and happy life. Larney, however, expresses a much more complicated ambivalence. On the one hand, like Alice Schafer, she questions whether her professors were really interested in her research potential. On the other hand, she recalls that she "didn't have any great fervor for research" and, in fact, seriously doubts whether women are naturally *capable* of research productivity in mathematics. "I still have a sneaking feeling that men are better at mathematical research than women," she says, citing as evidence the fact that women publish fewer papers and fewer total pages than men do (Larney: 11, 23).[2]

Both Larney and Schurrer launched their professional careers at state teachers colleges in the early 1950s. From the late 1950s onward, the mission, size, and scope of many of these institutions underwent radical transformation. When Schurrer first joined the faculty of Iowa State College for Teachers, she felt that her Ph.D. in mathematics distinguished her from her colleagues—who were mathematics *educators*—and made her a mathematician. But as her institution gradually metamorphosed into the University of Northern Iowa, her sense of professional identity changed as well, from "mathematician," to "mathematician *cum* teacher," and finally, simply, to "teacher" (Schurrer: 27–28). Like the research women, Schurrer equates being a mathematician with activity in research. The further removed one is from research and the more one finds oneself surrounded by those who are actively engaged in doing it, the less claim one has to the title "mathematician."

While Schurrer weathered the changes in her institution and her identity with equanimity, the transformation of the New York State College for Teachers into SUNY Albany was a traumatic one for Violet Larney. The academic year 1966–67 marked a turning point in the history of the mathematics department at Albany. In that year, the department added several new research faculty and moved to larger quarters on a new campus. Although Larney had done no original research beyond her dissertation, she had established herself as a pivotal figure in the department during the years prior to 1966, teaching a wide range of graduate courses and exercising leadership in the teacher education programs. Thus Larney was chosen to serve as acting department chair during the critical year of transition.

Larney speculates that if she had been active as a research mathematician, they might have given her the chair permanently. "The year I was chair," she recalls, "I met all these people, and set up the department programs, and did, well, everything that a chairman does and more." In particular, Larney presided over the expansion of the mathematics faculty from "about eight or nine" members to nearly twenty-five (Larney: 18).

After the critical year of transition, a permanent chair was appointed, and Larney stepped down from her leadership role. Under the new, research-oriented regime, "the first couple of years were really very bad" (Larney: 19). At first it appeared that she was going to be relegated to teaching nothing but lower-division service courses; but she stood her ground, ultimately carving out a niche for herself in the new department and gaining some recognition and appreciation for her efforts. "They made me the undergraduate chairman, in charge of setting up all the undergraduate courses and assigning people to teach them, and meeting with the students and helping plan the programs," Larney recalls. "I mean, I had some responsibility. I wasn't shunted to the corner to the extent that some of the others were" (Larney: 20). Even so, maintaining a sense of belonging in this highly territorial new environment was an ongoing struggle for Violet Larney, who felt as if she continually had to prove her worth to the research faculty.

With this struggle in mind, she expresses regret that she didn't pursue a different field of study—"possibly psychology"—in which the standards for research might have been less rigorous, where she might have made a contribution. "I guess, in my low moments," Larney says, "I thought, there's so much bull put out in those other fields, I could have published reams, and they would have called it 'research.' . . . And I would have gotten further ahead in some other field" (Larney: 20). In the final analysis, Larney believes—not without bitterness—that it is research that matters most in the contemporary American university: "When they say they're going to evaluate a faculty member . . . on their teaching and their research and their service to the university and the college, everybody knows which is the one that *really* counts. The others don't amount to a hill of beans if you're wonderful in research" (Larney: 12)

Violet Larney retired from SUNY Albany in 1982 and moved with her husband to Leisure World, a retirement community outside Phoenix, Arizona, shortly afterward. In retirement, her main preoccupations have included music, bridge, and travel. At Leisure World, when residents are getting acquainted they often ask one another, "What did you do in your other life?" Larney is generally reluctant to talk to her neighbors about her past involvement in mathematics, which really *does* seem to her as if it happened in another lifetime (Larney: 22).

Marie Wurster spent her entire professional career at Temple University in Philadelphia, which underwent an expansion and transformation comparable to that of the state teachers colleges during the forties, fifties, and sixties. Like Schurrer and Larney, Wurster readily acknowledges the tension between research and teaching; but unlike them, she proudly affirms her identity as a mathematician, an identity she has had since she earned the Ph.D. Wurster came to Temple University in 1946, at the height of the postwar enrollment surge, as the result of an unconventional job search. While completing her Ph.D. at Chicago in early 1946,

I read in [a news story in my hometown Philadelphia paper] that Temple was starting a special semester for World War II veterans, so I wrote applying, thinking that would tide me over until September. And then I received a phone call from the department chairman saying that they would like me to come, and would I be interested in a permanent position? Since I was happy to come back to my home in Philadelphia, I said that I would do so. So, really, once I had that offer—and financially it was a little better than anything else I had been offered—I took it and stopped looking. And so I often tell people, I started out with the idea of a job for six months, and I stayed for thirty-nine years! (Wurster: 10)

When she arrived at Temple, the mathematics department had just instituted a master's degree program. Wurster taught five three-hour courses per semester, at all academic levels, to meet the needs of a diverse and growing student population. "I had the opportunity to introduce a good number of advanced undergraduate and graduate courses—which, when you're young and enthusiastic, is very nice," Wurster recalls. She also taught her share of freshman-level courses "for business school students, or people who were just taking enough math to satisfy requirements" (Wurster: 11).

During those first hectic years at Temple, Wurster might have continued in research but for two obstacles: there was no one on the faculty to collaborate with, and teaching occupied nearly all of her time. She read widely in mathematics, taught a dizzying array of graduate and undergraduate courses, and eventually wrote and published three textbooks. Moreover, throughout the 1960s she held an NSF grant that supported educational programs for in-service teachers. Overall, she seems content with her commitment to teaching and educational outreach. She is somewhat regretful—but basically unperturbed—by the fact that she did no publishable research beyond the dissertation.

When Temple began its Ph.D. program in mathematics in the 1960s, her popularity, particularly in the graduate classroom, continued unabated. Harriet Lord (Ph.D., Temple, 1973) recalls that the female graduate students at Temple looked to Marie Wurster as an inspiration and a role model as they worked toward their own Ph.D.'s.[3] Yet despite her diverse contributions to the department, Wurster is aware that she was paid quite a bit less than the research faculty. "I think at one time the chairman of the math department told me that my salary was relatively cheap," Wurster recalls. "I had the impression, that maybe later in my career, especially when the department got larger and there were more research people and so on, that other people made a good bit more money than I did" (Wurster: 16). When the AAUP began to serve as a bargaining unit for the Temple faculty, Wurster finally received "a sizeable increase in salary," as redress for "past discrimination against women," and in 1984—two years prior to her retirement—Wurster finally achieved promotion to full professor (Wurster: 16).[4]

For Violet Larney, Augusta Schurrer, and Marie Wurster, the conflict between research and teaching was something that was mainly external to themselves: they worked in institutions whose values changed over time, in which researchers garnered an increasing share of rewards and recognition. For Barbara Beechler, the conflict between research and teaching was a much more internalized affair. As we have seen in chapter 5, Beechler was disheartened by her graduate research experiences. Of her dissertation, she says:

I found the whole thing horrid, and I swore that I would never write a paper and that I would never give a paper and that I would never publish anything. And I have managed to do that. I have never published. Not a word. I have worked on papers with other people, and so on, but I have always insisted that it's their paper. I just decided that. (Beechler: 31)

Although she had sworn off doing research herself, she never lost her zest for learning new mathematics, for being close to the forefront of mathematical research. Her experiences at Wilson College in the 1950s convinced her that she desperately needed the company of research mathematicians to feel connected to a living, breathing subject. While on the Wheaton College faculty in the 1960s, she participated in NSF summer institutes for talented high school students; she attended a summer institute in noncommutative algebra at Bowdoin College; and she took a year's sabbatical in Berkeley. She even went so far as to return to Iowa for a summer to participate in seminars run by her dissertation adviser, H.T. Muhly. She continually sought contact with creative mathematicians and in turn tried to convey the excitement of the field to her students.

"I kind of kept myself occupied and happy and off the streets," Beechler says of her years at Wheaton. But Wheaton was simply too isolating, too "closed-in," too limiting, both socially and intellectually (Beechler: 36). When Muhly died suddenly in December of 1966, she despaired of finding another job without his recommendation. But within a matter of months, she was offered a position at Pitzer College, the newest of the Claremont Colleges in California, and she jumped at the chance.

Pitzer was a women's college with strong emphasis in the social sciences; as such, it did not have a strong mathematics faculty. But the other Claremont Colleges, especially Pomona and Harvey Mudd, had research faculties, and the soon-to-be-established Claremont Graduate School promised to be a center of mathematical research and graduate training.

The move to Pitzer revitalized Barbara Beechler's life and career. In Claremont, she found—for the first time since her predoctoral years at Iowa and Smith—a genuine community of mathematicians, in which she could once again feel the excitement that first drew her to the subject. In gratitude for this sense of community, she made yet another vow that helped to heal her own unhappy relationship to research. "I decided that I would serve researchers," Beechler says. "I didn't care what kind of dirty jobs they made me do, I would do them, so that these guys could be free." The services she provided for her research colleagues ranged from dinners and cocktail parties to serving as "the founding mother" of the Claremont Mathematics Colloquium, to which she is so firmly devoted that she has included it in her will (Beechler: 42–43, 49).

Beechler's service to the research community in Claremont has not gone unrecognized or unrewarded. Although she retired in 1989, she remains active in the life of the colleges and has won the respect and affection of the Claremont mathematical community. "I'm not the worst mathematician in Claremont," she asserts with characteristically self-deprecating humor. "I have colleagues who are mathematical snobs of incredible dimension, and somehow or other they find excuses for me" (Beechler: 51).

Scholar-Teachers: Creativity or Compromise?

Barbara Beechler's attitude toward research brings her very close to the group of women I have identified in chapter 2 as *scholar-teachers*. Scholar-teachers acknowledge, and exemplify through their own work, the frequently seamless interdependence between good teaching and good scholarship, whether in the form of traditional mathematical research and expository writing or other, nontraditional forms of creative endeavor. The professional life of the scholar-teacher bears a striking similarity to the broad scholarship and teaching of many prewar American mathematicians. At the same time, the female scholar-teachers of the postwar era struck out in new directions, responding to the unique challenges and circumstances of the 1950s, 1960s, and beyond.

The careers of Winifred Asprey, Grace Bates, and Edith Luchins best exemplify the ethos of the scholar-teacher. Each of these women made professional contributions to emerging fields of scholarship—computing (Asprey), probability (Bates), and psychological aspects of mathematics education (Luchins)—and each one brought these contributions to bear on her work as a teacher. We have already seen in chapter 6 how Winifred Asprey, inspired by an earlier generation of Vassar mathematicians, ventured out into the emerging field of computer science and established computing facilities and a computing curriculum at Vassar.

Like Winifred Asprey, Grace Bates made some of her most significant scholarly contributions far afield from the area of her doctoral dissertation. When Bates joined the faculty of Mount Holyoke in 1946, she continued to pursue the research in algebra that she had begun under Reinhold Baer at Illinois. At the same time, however, she found herself teaching courses that went well beyond her area of research expertise. In those days, "it was a tradition everywhere that the statistics-probability course was given to the youngest person," Bates recalls, because "nobody wanted to teach it" (Bates: 16). As the junior member of the department, Bates inherited the course from the eminent Polish analyst Antoni Zygmund, who left Mount Holyoke for the University of Pennsylvania in 1945.

Bates "worked like a dog" in preparing the probability and statistics course and discovered to her surprise that she found the material fascinating. A serendipitous encounter with Zygmund led Bates to a startling change in research direction:

[Zygmund] came back for a social occasion [some] time later. And I was telling him how interested I was getting in the subject, but I really needed more education in it. And he said, "Well, I'll write to my friend out in California, Jerzy Neyman. I think maybe he could help you." And he apparently did, and I got this letter offering me an assistantship for the summer session there at the University of California at Berkeley. (Bates: 16–17)[5]

On Zygmund's recommendation, Bates worked for several summers in the 1950s with Neyman at Berkeley, coauthoring a number of papers in probability and statistics with him. In the 1960s she turned to the writing of elementary instructional materials on probability. While Bates's primary professional activity was teaching, she found over the course of her career that her experiences in the classroom led her into research and back again.

As we have seen in chapter 5, Edith Luchins began to collaborate with her husband in the study of psychological issues in mathematics education long before she returned to graduate school to complete the Ph.D. By the time she finally earned her doctorate in 1957, they had already produced several joint publications in the field. Luchins has published some research in the area of her doctoral dissertation—functional analysis, a branch of pure mathematics. But over the years her primary scholarly work has had a distinctly interdisciplinary flavor, combining the areas of psychology, mathematics, education, history, and gender issues. In this regard, she includes among her greatest mentors and role models the Gestalt psychologist Max Wertheimer, who was her husband's Ph.D. adviser and whom she met during her undergraduate years, and the historian of mathematics Carl Boyer, who was her teacher at Brooklyn College.

Given Luchins's lifelong interest in teaching, it is not surprising that she describes herself, first and foremost, as "an educator" (Luchins: 32). At the same time, she does not hesitate to identify herself as a mathematician; but what makes her a mathematician is a rather broad sense of engagement with the subject and with the profession. For Luchins, being a mathematician means "being versed, knowledgeable, in some aspects of the field" while at the same time "knowing my deficiencies" and "being eager to learn more." In other words, being a mathematician entails being a lifelong *student* of mathematics, while at the same time making—or *having* made—at least a modest contribution to mathematical research. Beyond this, she adds, it is important to remain "interested in the profession" and aware of "what's happening," from the employment situation for new Ph.D.'s to the status of Fermat's Last Theorem (Luchins: 31).

Herta Freitag's experience as a scholar-teacher has been informed by her conviction that mathematics is a part of culture and shaped by her awareness of the central role that problem solving plays in mathematics education. As a graduate student at Columbia, she was profoundly influenced by the mathematician Edward Kasner, who posed challenging, frequently unsolved, problems for homework.[6] Indeed, one of Freitag's first publications (Freitag and Freitag 1956) grew out of a problem Kasner posed in one of his graduate classes.

During her early years at Hollins College, Freitag was so busy designing curricula, teaching courses, and working with students that she found little time for problem solving and publication. In the early 1960s, however, she became first a subscriber and then a regular contributor to the *Fibonacci Quarterly,* a journal devoted to the study of sequences of numbers that arise, as do the Fibonacci numbers, from a simple set of recurrence relations. Over the past thirty-five years, Freitag has compiled an extensive publication list in what she refers to as "problem-posing, problem-solving" mathematics (Freitag-3: 7). In the years since her retirement from Hollins in the early 1970s, her activity has intensified; indeed, the November 1996 issue of the *Fibonacci Quarterly* (Volume 34, Number 5) is dedicated to her, in honor of her eighty-eighth birthday.

She characterizes her preoccupation with number-theoretic problems as "something that one does because one almost has to, as it's an urge that lies in the soul of somebody who is really dedicated to mathematics" (Freitag-3: 7). At the same time, she is careful to distinguish her work from "real research"—*forschung,* in German—which is the work of "top mathematicians," such as "Gauss and Archimedes and Newton," who reside "somewhere near heaven." In Freitag's view, her own humble contributions to mathematics do not meet her two criteria—mathematical significance and novelty—for genuine research. "I don't really consider it significant," she says, "because it does not add anything to the body of mathematics proper" (Freitag-3: 6–7). Yet despite her insistence that she works only on "little problems," some of her number-theoretic work has recently found application in cryptanalysis (Freitag-3: 10).

Whether it qualifies as "real research" or not, Freitag's enthusiasm for number theory and number relationships has always had a place in the classroom. Because the Fibonacci numbers are particularly easy to describe to a mathematically unsophisticated audience, she has been able to share her interest in them with generations of students—most recently, with students in the master of arts in liberal studies program at Hollins, where she taught a popular course in mathematics on a number of occasions during the 1990s. Following the example of her teacher Edward Kasner, Freitag believes that an interest in mathematics can be aroused in nearly every thoughtful person, once they are given the opportunity to experience for themselves the joy of discovering something new.

For Asprey, Bates, Luchins, and Freitag, the role of scholar-teacher has been a source of satisfaction and contentment and has brought them recognition and respect. But for other women of their generation, a synthesis of teaching and scholarship did not bring quite the same happiness or sense of fulfillment and did not fully resolve their conflicted feelings about mathematical research. For example, despite her exceptional early promise, Margaret Willerding found the male-dominated research community alienating and returned to the field of mathematics education whence she had come. As we have seen in chapter 6, however, her productivity and success in mathematics education have not erased the disappointment of her early years in research.

Similarly, it is easy to imagine that Rebekka Struik and Anneli Lax might have had very productive mathematical research careers under somewhat different personal and professional circumstances. Rebekka Struik published over a dozen papers in group theory during the first thirty years of her career, but for at least the first fifteen of those she was geographically isolated from other researchers in her field and had very little institutional support for her research program. As a result, she has often felt like an outsider in the community of algebraists; communication and collaboration with colleagues have often been very difficult for her. Over the past fifteen years or so, she has moved away from research in pure mathematics and has become increasingly preoccupied with mathematical activities that bear the hallmark of the scholar-teacher. Struik has taught a succession of advanced courses in fields such as history of mathematics, numerical analysis, and probability, in which she has had no previous training, using these courses as an opportunity to educate herself in areas of mathematics that are new to her. She has done a modest amount of research in the history of mathematics, joining a like-minded community of educators who are convinced of the importance of history to the teaching and learning of mathematics. In this respect, Rebekka

Struik is following in the footsteps of her father, Dirk Struik, for whom the history of mathematics has been a major preoccupation for the past fifty years (Rowe 1994).

Anneli Lax turned away from traditional mathematical research not long after earning the Ph.D. But from the early 1960s onward, she found a happy marriage of her dual interests in language and mathematics as editor of the New Mathematical Library, a series of high-quality expository texts, designed to make diverse areas of advanced mathematics accessible to a general mathematical audience (Albers 1992). At the same time, her work as an undergraduate teacher at NYU rekindled her longstanding interest in how people learn. As a teacher, Lax—who always had a keen awareness of her own learning process—was extremely "interested in why people found certain things difficult" and in "the stages at which people *do* understand something" in mathematics and come to a sense of satisfaction in this knowledge (Lax: 17). She was particularly interested in the role of language, and specifically writing, in the learning of mathematics. In collaboration with the creative writer Erika Duncan, Lax created an innovative course in mathematical thinking—"Mathink"—which has been successfully adapted for use in the public schools of Brooklyn, New York, with the support and encouragement of the Ford Foundation (Duncan 1989; Lax 1989).

Like Willerding and Struik, Lax at times felt alienated from the culture of mathematical research. Her interest in mathematics education, and in the problems of the New York City schools, brought her face to face with the "animosity between educators and academics." Her old friends and colleagues at the Courant Institute "had nothing *against* educating youngsters—but they didn't know how, or they wanted to leave that to other people." While many of them were interested in teaching "precocious children," Lax says that "the unwashed masses didn't interest anybody! But they *did* interest me." Lax characterizes the chasm between research mathematicians and mathematics educators as being "almost a political issue" (Lax: 17).

Lax expresses regret that she did not do more in the way of mathematical research, but her regret is tinged with a sense that the culture of research was too limiting, too unsupportive of her goals and needs as a person. "Had I done more . . . research, I feel, given my personality, [that]

I would have missed a lot of human stuff," she says. At the same time, however, she also believes that "at critical times, I didn't get the kind of support that I would have needed to 'fulfill my promise'" (Lax: 20). Like Willerding and Struik, Lax combined a strong commitment to teaching with the pursuit of more general, more interdisciplinary forms of scholarship in an effort to transcend the personal constraints and limitations of the postwar mathematical research ideal.

Activism and Opportunity: New Roles for Women Mathematicians

More often than not, it was a sense of limitations that inspired the women of this generation to strike out in new directions and thus to expand prevailing notions of what it means to be a mathematician. Winifred Asprey's move into computing was a means of escaping her midcareer rut. Before the major advances of the civil rights movement, Evelyn Granville found economic and intellectual opportunity in the federal government and at IBM. Margaret Martin and Joan Rosenblatt, sensing early on that academic mathematics was not for them, went on to uniquely satisfying careers in the statistical sciences. Unable to find a niche in the academic community of greater Boston, Vera Pless stumbled on an interdisciplinary community at the Air Force Cambridge Research Lab that led her into an exciting new field in mathematics, algebraic coding theory.

In the sixties and seventies, many of these same women took the lead in women's struggle for rights and recognition in the academic and mathematical communities. During the late 1960s, Alice Schafer helped to organize a series of "meetings of women mathematicians and mathematics graduate students in the Boston area to discuss common problems and possible solutions" (Schafer: 26). The core group of participants consisted primarily of women graduate students and recent (1960s) Ph.D.'s, but Alice Schafer took a prominent leadership role, and the group also included Kay Whitehead (Ph.D., Michigan, 1946), then an assistant professor at Tufts. At about the same time and also in the Boston area, Vera Pless joined forces with two women from the physics department at MIT, Vera Kistiakowsky and Elizabeth Baranger, to found the association that later came to be known as WISE (Women in Science and Engineering). There was significant overlap between the two organizations.[7]

By 1971, the Boston-area mathematicians' group had expanded to include women mathematicians from all over the United States and was organized into the Association for Women in Mathematics (AWM). Mary Wheat Gray (Ph.D., Kansas, 1964) of American University was the first president of AWM, and Alice Schafer succeeded her the following year. The fledgling AWM conducted consciousness-raising sessions at national mathematics meetings, advocating for gender equity in graduate education and academic employment. More generally, AWM sought solutions to the problems of women in mathematics, specifically, that of balancing work, marriage, and motherhood. While conditions for women in academic mathematics have generally improved since the early 1970s, the AWM continues to work toward the resolution of these same issues nearly thirty years later.[8]

The AWM brought enormous pressure to bear on the American mathematical community and on the AMS in particular. In response to this pressure, the AMS established its own Committee on Women in 1972, with Jane Cronin Scanlon among its founding members and Cathleen Morawetz as its first chair. Originally envisioned as a temporary committee with a short-term mission, the Committee on Women has expanded over the years to include representation from several other mathematical and statistical societies and continues to offer advice and recommendations for improving the status of women in mathematics.[9]

The struggle for gender equity in the academic community led to expanded opportunities for women in the 1970s and 1980s. Women were suddenly in demand for service on national committees and editorial boards; they were finally invited to give prestigious lectures at national and regional meetings, and received long-overdue promotions and tenure at colleges and universities. More than any other woman interviewed for this study, Lida Barrett has made the most of these expanded opportunities. In so doing, she transformed her life, her career, and her outlook on mathematics in ways that few women of her generation have done.

As we have seen in chapter 5, Barrett's graduate school years were marred by repeated experiences of discriminatory antinepotism. When at last she earned the Ph.D. in 1954, the fifteen-year period that followed was marked by personal and professional adversity. After sustained attempts to start a family in the early 1950s, Lida and John Barrett learned that they were unable to have children and so began a long and exasperating struggle to become adoptive parents. Passing up promising early job prospects at Wellesley and Bryn Mawr, Lida Barrett decided that her husband's career track would take precedence.[10] Like so many other women of her generation, her primary commitment was to her family; beyond that, she would find the best mathematical job available, wherever she happened to be living.

During the years 1956 to 1961, the Barretts lived in Utah, where they successfully adopted three children. While her husband held a regular professorship at the University of Utah, Lida Barrett remained at the rank of lecturer, despite her continuing research in general topology, the subject area of her dissertation. In 1961, they moved to the University of Tennessee, with a view to being somewhat closer to their families and hopeful that Lida might find better job opportunities. At Tennessee, however, antinepotism rules prevented her from holding a regular tenure-track job. Although she had the normal workload of a tenure-track professor—teaching two courses, pursuing her own research, supervising theses and dissertations—she was employed in a succession of one-year appointments, as an "acting associate professor, part-time" (Barrett: 19). Eventually, she took a year off from the university, during which she stayed home with the children and started a small but successful consulting business, with Oak Ridge National Laboratory among her chief clients.

While Lida Barrett was clearly displeased with her second-class citizenship during those early years at Utah and Tennessee, professional problems were among the least of her concerns. Her husband had been diagnosed with polycystic kidney disease, and during the sixties his health steadily deteriorated. In late 1968, his kidneys failed completely, and after a long-awaited kidney transplant was rejected, he died in early 1969. John Barrett's death, though not entirely unexpected, came swiftly and suddenly and dealt a tremendous blow to Lida Barrett and the children.

But this tragic loss also heralded the beginning of a period of growth and creativity for Barrett, who made the most of every opportunity that came her way. Indeed, such opportunities came with startling frequency. In 1969 she was offered a regular tenured slot in the mathematics department at Tennessee and became a full professor a year later. "I was the third female full professor," Barrett recalls, "outside of home ec. and women's P.E. and nursing" (Barrett: 28). Her status as one of the very few women full professors, combined with her considerable social and organizational skills, made her a popular choice for university and college committees. In 1972, she was asked to chair a special task force on women at the University of Tennessee that was charged with investigating the climate for women on campus. The task force made substantive recommendations regarding salary equity, child care, equal access to recreational facilities, and use of gender-inclusive language. Not long thereafter, Barrett was offered a position as associate dean.

As a single parent of three children, with research and consulting work to be done, she declined the offer. But a move into academic administration seemed all but inevitable: universities were under mounting legal pressure to put women in positions of leadership, and Barrett's talents and skills were hard to overlook. In 1973, she was invited to apply for the job of mathematics department head at the University of Alabama. When Alabama offered her the job, Tennessee made a counteroffer, which she decided to accept. From 1973 to 1980, Barrett was the head of the Tennessee math department—one of the first women in the country to hold such a post at a major university.[11]

At first, Barrett says, "I didn't realize how much authority I really *had*. I'd always been the wife and the younger sister, and I had been impatient about all the things that needed to be done" (Barrett: 25). But she quickly learned how to use her newfound power. Under Barrett's leadership, the department added an applied mathematics program; faculty members not significantly involved in research were given higher teaching loads and encouraged to improve their teaching; and the national ranking of the department increased substantially. Barrett proved to be an effective administrator, winning the trust and respect of her colleagues.

During the academic year 1979–80, she took part in a program sponsored by the American Council on Education (ACE) to identify senior women faculty members who might have an interest in higher education administration. Her involvement in the ACE program led to an offer from Northern Illinois University, which was looking for an associate provost for undergraduate education who could help oversee the addition of a new program in engineering. After several years as department head, and with all three children in college, Barrett felt ready for a change and accepted the position at Northern.

When Barrett became head of the department at Tennessee, her career in research effectively ended; with the move to Northern, she moved still further from her roots as a research mathematician. In her work as associate provost, curricular matters were always at the forefront of her concerns. Thus it was quite natural that her interest in curriculum and her background in mathematics should come together, and during the 1980s she became increasingly involved in the calculus reform movement.

In 1987, she became dean of arts and sciences and professor of mathematics and statistics at Mississippi State University, a move made in part to bring her closer to the day-to-day issues of college teaching. While she continued to teach mathematics both at Northern and at Mississippi State, her primary involvement with mathematics was at the national level, through the AMS and the MAA. After several years of service on important committees in both organizations, she was elected to a two-year term as president of the MAA, only the second woman to hold the office.[12] Her interest in and involvement with matters of national policy in the sciences and mathematics led naturally to an appointment as senior associate at the National Science Foundation in 1991. She remained at the NSF until 1995 when, with retirement in view, she returned to college teaching at the United States Military Academy at West Point.

The trajectory of Lida Barrett's adult life has been anything but simple. She learned early on that it was futile to expect her life and career to conform to some preconceived notion of the life course because so much of what has happened—for good and for ill—has been unexpected and beyond her control. "I never intended to be a widow at forty-one," she recalls, but like her father before her, she transformed adversity into opportunity (Barrett: 33). Continually responding to the challenges of a complex personal and professional environment, Barrett has inhabited a variety of roles—wife, mother, widow; researcher, professor, consultant; administrator, policymaker, association president. And yet, she maintains a clear, coherent, and very broad sense of herself as a mathematician. She first began to think of herself as a mathematician while an undergraduate student at Rice. True to her rich and varied experience, she says, "I hope that people can think of mathematicians in a huge variety of roles" (Barrett: 37).

The Question of Identity

Thus we return to the question that opened this chapter: what does it mean to be a mathematician? Some of the interviewees began to think of themselves as mathematicians at a fairly early age; others didn't have that sense of themselves until much later in life. But nearly all would agree on one thing: earning the Ph.D.—tackling an unsolved mathematical problem, struggling with it, solving it—was crucial to the process. The experience of having created new knowledge in mathematics, if only once in one's life, establishes a lasting connection to the subject. Even Jean Walton, whose predilections led her away from academic mathematics and into a very different life as a dean, acknowledges that it took nearly twenty years after the Ph.D. for her to sever her last connection to the field.

As we have seen, many of the interviewees take the narrow view that one cannot truly be a mathematician unless one is engaged in research. Certainly, research must play a central role in any living, growing subject. But the "variety of roles" that Lida Barrett speaks of make for a variety of ways in which both women and men can be a part of contemporary mathematics while still living full and satisfying lives. Anneli Lax agrees that there are many different styles, approaches, and ways-of-being in the mathematical world. She says that, for her, being a mathematician has to do with "the way I look at the world, more than what I have achieved." She is a mathematician because of certain "analytical, critical" habits of mind that have come from working for many years with mathematical language and mathematical ideas (Lax: 19–20). For Lax, these habits of mind found application not only in mathematical research but in teaching students to write and think clearly about mathematics and the larger world in which they live.

Nearly all the interviewees acknowledge that involvement with a community of people who care about mathematics is crucial to self-identification as a mathematician. Anneli Lax's sense of herself as a mathematician was shaped and molded through her marriage to Peter Lax and her long affiliation with the community at Courant Institute. Maria Steinberg, who has done no research since the early fifties and who retired early from teaching in the mid-1970s, nevertheless feels connected to the subject through her husband and their many mutual friends in mathematics. "I've always been a mathematician," Steinberg says, adding that she *continues* to be one "because I also *live* among mathematicians" (Steinberg: 18). But Jean Walton, who was isolated from the community of mathematicians as a graduate student, never developed a sense of how important social connections in mathematics really are.

Lida Barrett's experience shows that the work of a mathematician can go far beyond a narrow focus on research to embrace teaching, administration, community leadership, and public policy. Moreover, her experiences and those of many other women in this group show conclusively that one can be a mathematician while at the same time maintaining many other interests and values, as well as crucial connections to family and community. Barbara Beechler perhaps expresses it best when she says that being a mathematician means that one "keeps on worrying about mathematics" throughout one's life (Beechler: 50). In a variety of ways, on a variety of timetables, in lives punctuated with surprises and tragedies, these women have worried and cared about mathematics for over half a century. In so doing, each has rightfully earned the title of *mathematician.*

8

Dimensions of Personal and Professional Success

Five Key Elements of Work and Life Success

By most conventional measures—tenured professorships, publications, professional honors, influence on generations of students—the thirty-six women interviewed for this study have been extraordinarily successful. But as they look back over their lives, what does success really consist of for this group of women mathematicians? On careful study of the interviews in their entirety, it seems possible to identify at least five major factors that contribute to a sense of success in work and life among the women of this group.

First of all, one must have adequate opportunity to explore one's interests and talents and to determine which of these one most wants to pursue and develop. The myth of the mathematical life course clearly suggests that mathematics is a vocation to which only a few innately talented individuals are called. But as we have seen in earlier chapters, many of the interviewees chose mathematics from among a wide range of educational and professional options and can easily imagine having done something different. For example, Tilla Weinstein says, "In the greedy sense that we would all like to live many different lives and enjoy many different experiences, I would love to have done many different things," including musical composition, creative writing, and work in the apparel industry (Weinstein: 42). Because her mother encouraged her as a young child to imaginatively explore her interests and options, however, Weinstein feels that she was not forced into any particular field but made a conscious choice to pursue mathematics.

Patricia Eberlein is unique among the interviewees in that she actually experimented with many of the life and career options that Tilla Weinstein merely dreamed about. In the process, Eberlein discovered that, in fact, mathematics was what she loved. Perhaps because she took the time for such active exploration, she was able to take a flexible approach to her subsequent career in mathematics. Eberlein, who worked for several years in academic computing and later became a professor of mathematics and computer science, found contentment in a variety of roles and circumstances.

The second major component of success among the interviewees is the ability, by means of a combination of education and employment, to develop one's talents through creative and meaningful work. Winifred Asprey's experience illustrates how, within the context of a single job, it is possible to continue learning and growing and even to change one's career. Asprey's shift from pure mathematics and mathematics teaching into computer science and academic computing provided her a means of continuing her education throughout life, using a wide array of intellectual and personal skills and talents along the way. In a similar vein, the choices made by Jean Walton as she became increasingly committed to her work as dean at Pomona College ultimately led her away from mathematics entirely. Even so, Walton believes that completing the Ph.D. under Hans Rademacher at Penn contributed immeasurably to her sense of mastery and accomplishment, giving her the courage and confidence to achieve many other things later in life. Her early involvement with mathematics was an important stage in her personal process of discernment, and as such, she does not regret it.

A third critical element for success among the interviewees is a network of significant connections with others: the enjoyment of friendship, intimacy, and family life as well as a sense of belonging to a larger community of shared interests and values. As we have seen, many of these women married and had children, and some had unusually close intellectual partnerships with their husbands. Edith Luchins, for example, has been a lifelong collaborator with her husband on projects in psychology, education, and history. She describes her five children as her proudest accomplishment. At the same time, her life has been grounded in and enriched by her Jewish faith and her deep involvement with the Jewish community.

Her spouse, her family, and her community have been the touchstones of all that she has achieved.

For those women who did not marry, involvement in a wider community has been all the more critically important. For Jean Walton, who as a graduate student was almost completely isolated from the communal aspects of mathematical life, what made her work as dean most thoroughly satisfying was the way in which it completely engaged her in the day-to-day life of a college and its students, first at Swarthmore and later at Pomona. For Barbara Beechler, the community of mathematicians at the Claremont Colleges has energized and sustained her, much as the community of Vassar College has done for Winifred Asprey over the years.

The opportunity to enjoy the recognition and acclaim of others for what one has accomplished constitutes a fourth key ingredient of work and life success. Promotion, tenure, invitations to speak at major conferences, appointments to important committees and editorial boards, and awards and prizes are just a few of the outward signs of such recognition. For several of the interviewees, special acknowledgment by their institutions has been especially meaningful. At Rensselaer Polytechnic Institute, where she taught mathematics for over thirty years, Edith Luchins was named "a woman who has made history at the Institute," and her portrait now hangs in the RPI library (Luchins: 38). Vassar has named a computing lab on campus after Winifred Asprey to salute her pioneering role in the establishment of computing at the college. The Association for Women in Mathematics honored Alice Schafer in 1989 by establishing "the Alice T. Schafer Mathematics Prize, to be awarded annually beginning in 1990 to an undergraduate woman for excellence in mathematics" (Schafer: 29). Such awards provide independent validation of one's life and work.

At the same time, in the absence of such external affirmation, it is difficult to persevere in mathematics—or indeed, any other area of endeavor—with enthusiasm. As a graduate student at Penn, Jean Walton received little in the way of positive feedback on her progress from her adviser; and in her isolation from her fellow students and from the larger mathematical community, she found little encouragement to go on in mathematical research—particularly when her earlier work as a dean at Swarthmore had been so deeply satisfying and gratifying to her. It is perhaps a similar lack of professional recognition that leads Margaret

Willerding to say, as she looks back over her extraordinarily prolific career in mathematics education, "It was a lot of work, and I don't know whether it was worth it or not." When asked whether she would pursue a mathematical career, given the opportunity to live her life over again, she says, "I don't think I'd do it again" (Willerding: 12).

For the interviewees, the fifth and final key element of career and life success is the opportunity to "make a lasting contribution to future generations"—what the psychologist Erik Erikson has referred to as the expression of "generativity."[1] Such lasting contributions can be made through one's own children, of course, but also through research and (perhaps especially) through teaching. For Grace Bates, interacting with individual students and making a difference in their lives brought more "joy and exultation" than any other activity. "It's so much fun to get a letter from some student you had years ago, who still remembers you, and says so," says Bates. "Those are the rewards of a teacher, I think— not money" (Bates: 23).

Winifred Asprey still has an office at Vassar College, and when contemporary students meet her on campus for the first time, they ask if the Asprey Computing Lab is named for her father or husband. "When they find out it's [me], they look at [me] as if [I] should have been dead years ago! But they recover," she says. Eventually, they come to regard her as an ally and a resource: "I get to know them very well, and they use me in a way that they don't use the other members of the faculty because I'm not grading them." They ask her for help with personal problems, and they ask for her advice in the job search, too. "They figure that I know a lot more about jobs and have more connections around the country than do the younger faculty," Asprey says, "which I think is probably true." For Asprey, as for many other women of her generation, continued contact with young people is what "keeps [her] alive" (Asprey: 55).

Effort, Ability, and Luck

These, then, are five key attributes of a successful personal and professional life, as seen through the eyes of the women of the 1940s and 1950s generation of mathematics Ph.D.'s. What are the various factors to which these women attribute their success?

While success in work and life comes about by a combination of factors, people commonly attribute their successes (and failures) to just three: effort, ability, and luck (Valian 1988: 169–171). Of these three, it is *effort* that appears to be the one under the greatest conscious control, since plugging away at a task seems to be a matter of will and perseverance. *Ability,* while to some degree dependent on genetic factors, can also be developed through study and practice. *Luck* tends to be used as a catchall term that describes any and all forces that influence an individual's performance and are entirely outside her conscious control.

In her landmark study of women's comparatively slow progress in the professions, Virginia Valian (1998) observes that women tend to attribute their professional successes to luck, rather than to effort or ability, over which they have considerably greater control. Valian argues that the persistence of gender discrimination, particularly in male-dominated fields, means that women's effort and ability are only inconsistently rewarded and are sometimes met with indifference or disapproval. Because women frequently experience their professional successes as "due to random or uncontrollable factors," they find it difficult to take responsibility for success when it arrives (Valian 1998: 183).

Relying on luck as an explanation for either success or failure, however, is a professional liability for women:

If you see a success or failure as due to luck, you cannot learn anything from it. There is no point in trying to figure out what went wrong or right, no point in developing a plan for the future based on the past, no point in putting forth a lot of effort the next time. Chance undoubtedly enters into every result, but consistent success demands competence, strategic analysis, and effort. (Valian 1998: 21)

In light of Valian's observations, it is not surprising that many of the interviewees attribute their professional successes to external factors such as luck and timing. Augusta Schurrer provides an illustrative example. "I think my timing has been unusually good," she asserts. Had she not finished college in 1945, when there were still very few men in graduate school, she doubts that she would have qualified for graduate support at Wisconsin. Indeed, just a few years later, "you had to be really something *special . . .* to make it through or to have anybody be willing to work with you," she says, adding, "I wasn't that good" (Schurrer: 29–30). But the fact of the matter is, she *was* that good. She attracted the attention of the complex analyst Morris Marden, who was not only very invested in working with her as a graduate student but also tried to persuade her to return to research with him several years later. Schurrer is clearly pleased with her career as a teacher of mathematics at the University of Northern Iowa. But how might her life and career have been different had she come to claim the power inherent in her own ability and effort?

In Tilla Weinstein's account of her mathematical life, it is heartening to observe an evolving pattern of attribution as her career unfolded and she compiled an increasingly impressive record of achievement. She clearly regards her early achievements in college and graduate school as due to luck, timing, and "general intelligence" rather than her own mathematical ability. At NYU, she says, the faculty took an interest in her because there was a general shortage of mathematical talent. First they "needed" her to be the third member of the Putnam exam team, and then they "needed" her to fill a seat in graduate school because the country "needed" mathematics faculty. Of her admission to doctoral study, she says, "There was no sense that some terrible mistake would be made if you turned out someone who was not great" (Weinstein: 8–9). Thus Weinstein, like Schurrer, makes light of the connection between her early success and her basic ability.

But Weinstein attributes her postdoctoral research accomplishments to "hard work" (that is, to effort), and by 1980, she says that mathematics was "finally, flowing from inside of me" with the kind of effortless ease that most people describe as talent (Weinstein: 10, 19). Over time, as she developed a sense of competence, mastery, and control over her personal and professional environment, she became increasingly comfortable with taking credit for her own success.

While some of the interviewees displayed unusual confidence at a very early stage in their careers—Domina Spencer, Mary Ellen Rudin, and Edith Luchins come to mind—the majority of those who seem the most satisfied with what they have done in their lives and careers follow a similar pattern to Tilla Weinstein's. They develop a sense of competence, of personal efficacy and agency, through a gradual process that comes to fruition at about middle age. At the same time, in a parallel development, they come to understand that they are not solely responsible for all of their failures. Other factors—discrimination, injustice, caprice, timing, luck—play a larger role in their setbacks than they had previously imagined. Ultimately, they come to realize that, in work and in life, "You take the hand that's dealt you, you look at the challenges that are there, and you meet them," head on (Barrett: 36).

The Mathematical Life Course, Revisited

It is impossible to tell, from a study of this kind, just how many women of this generation either left mathematics, or never pursued it in the first place, because of external factors that they could not overcome. But as the stories of the interviewees make clear, it *is* possible for women to overcome considerable obstacles to their progress in mathematics. At the same time it is equally clear that the women mathematics Ph.D.'s of the forties and fifties enjoyed, albeit belatedly, some extraordinary opportunities. Many of them were able to reap the benefits of the women's movement at midlife, just as they were beginning to hit their stride mathematically. Many of them took advantage of opportunities for personal and professional growth that led them away from mathematics entirely and into new communities and new areas of intellectual endeavor.

The women mathematics Ph.D.'s of the forties and fifties lived in a turbulent time of social, political, and cultural growth and change. As we have seen, the rigid and inflexible model of career development put forward in the myth of the mathematical life course was never a viable model for them to follow. Indeed, changing times and circumstances require courage, flexibility, resilience, and the willingness to strike out in new directions. The experiences of the women mathematicians of the postwar generation provide a model for both men and women, in mathematics and in many other fields as well, who must face the challenge of building a meaningful personal and professional life at the turn of a new millennium.

"Being a mathematician has been very exciting for me," says Mary Ellen Rudin, "but I don't know that I had any vision of it as a twenty-year-old." In particular, the joy she has found in being a mathematician is "not something I could have anticipated" (Rudin: 28). What she shares with the most successful women mathematicians of this—and indeed, of *any*—generation is the dedication and the inventiveness to create a life centered around both work and people she loves and cares about.

Appendix A
Note on the Methodology of Oral History

This study is based on in-depth, tape-recorded interviews that were conducted during the period from October 1995 through August 1997 with thirty-six women who earned Ph.D.'s in mathematics from American institutions during the years 1940 through 1959. All but one of the interviews was conducted in person. My aim in each interview was to elicit a narrative account of the interviewee's life and career in mathematics. In all but the first few cases, each interviewee was asked to complete a biographical questionnaire in advance of our meeting. Once we finally met in person, I also asked them to provide supporting documents, including photographs, *curriculum vitae,* and other written materials.

Each interview was based on a list of twenty-one questions, proceeding more or less chronologically through the interviewee's family background, early education, college and graduate study, and career development. Special attention was paid to the interweaving of personal, family, and career concerns. The interviews themselves were somewhat open-ended. Although I worked in each case from the same set of questions, not all of the questions were actually *asked* of each interviewee, since many of them were able to anticipate the issues I wanted to discuss. Moreover, I endeavored to frame the questions in terms that were as neutral as possible, in an effort to bring out each woman's authentic point of view.

For a variety of reasons, some of the interviews were unusually brief, but typically, each interview went on for a period of one to three hours. The interviews were conducted following the procedural, legal, and ethical guidelines put forth by the Oral History Association. In particular, each interviewee gave her informed consent prior to the interview and signed a deed of gift afterward, transferring copyright of the interview material to me. The forms used in the interview process were approved for use in this project by the Institutional Review Board at Virginia Tech.

I prepared a verbatim transcript of each interview, and a copy of the transcript was sent to each interviewee for review and revision. At some future point, the tapes and transcripts will be made available to the public in an oral history archive. An alphabetical list of the interviewees, which includes the date and location of each interview and the length of the transcript in pages, is given in appendix B. Because four of the interview transcripts (Bell, Gassner, Hahn, and Morley) are not yet in their final, archival form, quotations from these interviews appear very rarely in the text.

Information gleaned from an interview narrative is highly subjective, reflecting an individual's reaction to and interpretation of life events. With a view to placing the narratives in proper perspective, I have done additional research using primary and secondary sources in an effort to verify the interviewees' factual accounts and to provide a historical, cultural, and mathematical context for their recalled experiences.

For the most part, the interview extracts presented in this book are presented as direct quotations from the interviewees, removed from the context of the interview itself. In particular, the questions that prompted these responses have generally been omitted. In the text, I have endeavored to supply the proper context for such quotations. Moreover, I am keenly aware of the fact that the interviewees and I may have differing views on the interpretation of circumstances and events. (Borland 1991 provides a thoughtful discussion of the problems inherent in attempts to interpret women's oral narratives of the past.)

My attitude throughout has been one of profound respect toward the women who have so generously shared their stories. At the same time, I believe that a retrospective view, informed by the insights of liberal feminist scholarship, can shed new light on the lives of twentieth-century women mathematicians.

As this book goes to press, four of the interviewees are no longer living and a number of others are in failing health. In the text, however, each woman's words are used as if she were speaking in the present, thus preserving a sense of the immediacy and spontaneity of the interview setting.

Appendix B
Alphabetical List of Interviewees

Each entry in the list contains the following information. The name of the interviewee appears first, followed in parentheses by the Ph.D. institution, year of Ph.D., date of birth, and sometimes date of death. Then the Ph.D. adviser and field of study are listed, followed by information on the interviewee's marital status and children (if any). If the individual held a tenured college or university position at some point in her career, this is noted next. Finally, the interview date(s) and location(s) are given, together with transcript length in pages.

A surname in (parentheses) denotes a married name under which the individual has been known at some time during her career; a surname in **boldface** denotes the name under which the individual has been most recently known professionally; a surname in *italics* denotes the name under which the Ph.D. was awarded.

Anne *Lewis* **Anderson** (Ph.D., Chicago, 1943; born 1919). Adviser: Magnus Hestenes. Field: Calculus of variations. Widowed, no children. Tenured. Interviews: by phone, 18 April 1995, Chapel Hill, N.C. (13 pages); in person, 20 November 1995, Chapel Hill, N.C. (8 pages).

Winifred *Asprey* (Ph.D., Iowa, 1945; born 1917). Adviser: E.W. Chittenden. Field: Analysis. Never married, no children. Tenured. Interview: 24 August 1996, Poughkeepsie, N.Y. (56 pages).

Lida Kittrell *Barrett* (Ph.D., Pennsylvania, 1954; born 1927). Adviser: J.R. Kline. Field: General topology. Widowed, three children. Tenured. Interview: 15 August 1997, West Point, N.Y. (37 pages).

Grace *Bates* (Ph.D., Illinois, 1946; 1914–1996). Adviser: Reinhold Baer. Field: Algebra. Never married, no children. Tenured. Interview: 13 June 1996, Newtown, Pa. (24 pages).

Barbara *Beechler* (Ph.D., Iowa, 1955; born 1928). Adviser: H.T. Muhly. Field: Algebra. Divorced, no children. Tenured. Interview: 23 May 1996, Claremont, Calif. (51 pages).

Janie *Lapsley* Bell (Ph.D., Illinois, 1943; born 1913). Adviser: E.B. Coble. Field: Geometry. Widowed, three children. Not tenured. Interview: 22 May 1997, Monument, Colo. (18 pages).

Anne *Whitney* Calloway (Ph.D., Pennsylvania, 1949; born 1921). Adviser: I.J. Schoenberg. Field: Complex analysis and approximation theory. Married, two children. Not tenured. Interview: 21 October 1996, Kalamazoo, Mich. (23 pages).

Mary Dean *Clement* (Ph.D., Chicago, 1943; born 1914). Adviser: E.P. Lane. Field: Geometry. Never married, no children. Not tenured. Interview: 18 November 1995, Nashville, Tenn. (40 pages).

Jane *Cronin* Scanlon (Ph.D., Michigan, 1949; born 1922). Adviser: E.H. Röthe. Field: Functional analysis. Divorced twice, four children. Tenured. Interview: 12 June 1996, Highland Park, N.J. (49 pages).

Patricia James (*Wells*) Eberlein (Ph.D., Michigan State, 1955; 1925–1998). Adviser: Charles P. Wells. Field: Applied mathematics, integral equations. Divorced twice, widowed, seven children (includes stepchildren). Tenured. Interview: 12 August 1997, Buffalo, N.Y. (18 pages).

Herta Taussig *Freitag* (Ph.D., Columbia, 1953; 1908–2000). Adviser: Howard Fehr. Field: History of mathematics, mathematics education. Widowed, no children. Tenured. Interviews: 28 October, 11 November, and 2 December 1995, Roanoke, Va. (23, 25, and 19 pages).

Betty Jane *Gassner* (Ph.D., NYU, 1957; born 1934). Adviser: Wilhelm Magnus. Field: Algebra. Never married, no children. Not tenured; not employed in academia. Interview: 7 June 1997, New York, N.Y. (8 pages).

Evelyn *Boyd* Granville (Ph.D., Yale, 1949; born 1924). Adviser: Einar Hille. Field: Analysis. Divorced, remarried, no children. Tenured. Interview: 19 May 1997, Tyler, Tex. (28 pages).

Susan Gerber *Hahn* (Ph.D., NYU, 1957; born 1914). Adviser: Peter Lax. Field: Partial differential equations. Widowed, no children. Not tenured; employed in industry. Interview: 6 June 1997, Queens, N.Y. (39 pages).

Violet *Hachmeister* Larney (Ph.D., Wisconsin, 1950; born 1920). Adviser: C.C. MacDuffee. Field: Algebra. Married, no children. Tenured. Interview: 21 May 1996, Mesa, Ariz. (24 pages).

Anneli Cahn *Lax* (Ph.D., NYU, 1955; 1922–1999). Adviser: Richard Courant. Field: Partial differential equations. Divorced, remarried, two children. Tenured. Interview: 5 June 1997, New York, N.Y. (21 pages).

Edith Hirsch *Luchins* (Ph.D., Oregon, 1957; born 1921). Adviser: Bertram Yood. Field: Functional analysis. Married, five children. Tenured. Interview: 14 August 1997, Troy, N.Y. (41 pages).

Dorothy *Maharam* Stone (Ph.D., Bryn Mawr, 1940; born 1917). Adviser: Anna Pell Wheeler. Field: Measure theory. Married, two children. Tenured. Interview: 28 August 1996, Boston, Mass. (15 pages).

Margaret *Owchar* **Marchand** (Ph.D., Minnesota, 1950; born 1925). Adviser: Robert Cameron. Field: Analysis. Divorced, three children. Tenured. Interview: 22 October 1996, Adrian, Mich. (25 pages).

Margaret *Martin* (Ph.D., Minnesota, 1944; born 1915). Adviser: Dunham Jackson. Field: Analysis. Never married, no children. Tenured; later, U.S. government employee. Interview: 21 July 1997, St. Paul, Minn. (28 pages).

Cathleen Synge *Morawetz* (Ph.D., NYU, 1951; born 1923). Adviser: Kurt O. Friedrichs. Field: Applied mathematics. Married, four children. Tenured. Interview: 4 June 1997, New York, N.Y. (28 pages).

Vivienne Brenner *Morley* (Ph.D., Chicago, 1956; born 1930). Adviser: Antoni Zygmund. Field: Harmonic analysis. Married, no children. Not tenured. Interview: 13 August 1997, Ithaca, N.Y. (25 pages).

Vera Stepen *Pless* (Ph.D., Northwestern, 1957; born 1931). Adviser: Alex Rosenberg. Field: Algebra. Divorced, three children. Tenured. Interview: 16 July 1996, Oak Park, Ill. (29 pages).

Joan Raup *Rosenblatt* (Ph.D., North Carolina, 1956; born 1926). Adviser: Wassily Hoeffding. Field: Mathematical statistics. Married, no children. Not tenured; U.S. government employee. Interview: 2 October 1996, Gaithersburg, Md. (23 pages).

Jean Hirsh *Rubin* (Ph.D., Stanford, 1955; born 1926). Adviser: C.C. McKinsey and Patrick Suppes. Field: Logic. Married, two children. Tenured. Interview: 15 July 1996, West Lafayette, Ind. (16 pages).

Mary Ellen *Estill* **Rudin** (Ph.D., Texas, 1949; born 1924). Adviser: R.L. Moore. Field: General topology. Married, four children. Tenured. Interview: 17 July 1996, Madison, Wis. (28 pages).

Alice *Turner* **Schafer** (Ph.D., Chicago, 1942; born 1915). Adviser: E.P. Lane. Field: Geometry. Married, two children. Tenured. Interview: 1 October 1996, Arlington, Va. (31 pages).

Augusta *Schurrer* (Ph.D., Wisconsin, 1952; born 1925). Adviser: Morris Marden. Field: Complex analysis. Never married, no children. Tenured. Interview: 20 July 1996, Cedar Falls, Iowa (34 pages).

Domina Eberle *Spencer* (Ph.D., MIT, 1942; born 1920). Adviser: Dirk J. Struik. Field: Tensor analysis. Widowed, one child. Tenured. Interview: 26 August 1996, Boston, Mass. (42 pages).

Maria *Weber* **Steinberg** (Ph.D., Cornell, 1949; born 1919). Adviser: William Feller. Field: Partial differential equations. Married, no children. Tenured. Interview: 25 May 1996, Pacific Palisades, Calif. (21 pages).

Ruth Rebekka *Struik* (Ph.D., NYU, 1955; born 1928). Adviser: Wilhelm Magnus. Field: Algebra. Divorced, three children. Tenured. Interview: 21 May 1997, Boulder, Colo. (39 pages).

Jean *Walton* (Ph.D., Pennsylvania, 1948; born 1914). Adviser: Hans Rademacher. Field: Analytic number theory. Never married, no children. Employed as college dean. Interview: 24 May 1996, Claremont, Calif. (37 pages).

Tilla Savanuck (*Klotz* Milnor) **Weinstein** (Ph.D., NYU, 1959; born 1934). Adviser: Lipman Bers. Field: Complex analysis and geometry. Divorced twice, remarried, two children. Tenured. Interview: 11 June 1996, Metuchen, N.J. (44 pages).

Margaret *Willerding* (Ph.D., St. Louis, 1947; born 1919). Adviser: Arnold Ross. Field: Number theory. Never married, no children. Tenured. Interview: 26 May 1996, La Mesa, Calif. (13 pages).

Joyce White *Williams* (Ph.D., Illinois, 1954; born 1929). Adviser: Waldemar Trjitzinsky. Field: Integral equations. Widowed, five children. Tenured. Interview: 27 August 1996, Lowell, Mass. (20 pages).

Marie *Wurster* (Ph.D., Chicago, 1946; born 1918). Adviser: Lawrence Graves. Field: Calculus of variations. Never married, no children. Tenured. Interview: 9 June 1997, Swarthmore, Pa. (20 pages).

Notes

Introduction

1. One measure of women's participation in scientific disciplines is given by the proportion of women among the Ph.D.'s awarded in each field. For detailed statistics on Ph.D.'s awarded in the United States, by field of specialization and gender, during the years 1920 to 1974, see Harmon 1978. Aggregated U.S. statistics for the years 1920 to 1989 are effectively presented and discussed in Stephan and Kassis 1997. For other measures of women's participation in science in America, see Rossiter 1982, 1995.

Chapter 1: Women Mathematicians and the World War II Transition

1. Sally Gregory Kohlstedt (1978) argues that nineteenth-century women in American science comprise three distinct generations: the "independents," who studied science with private support; the "educators and popularizers," who served as science teachers and authors of textbooks; and finally, the "professionals," who sought to join the scientific community and to achieve social recognition for their work. The participation of nineteenth-century women in mathematics in the West roughly follows this pattern.

2. Although the awarding of doctorates of various kinds dates back to the early medieval University of Bologna, the doctor of philosophy degree seems to have been a much later German creation (Simpson 1983). I have been unable to determine when and where the first Ph.D. in mathematics was awarded, but Carl Friedrich Gauss earned one in Germany as early as 1799. Early Ph.D.'s were frequently awarded by universities for work completed elsewhere; Gauss, for example, was a student at the University of Göttingen but submitted his research and was awarded the Ph.D. *in absentia* at the university at Helmstedt (Hall 1970: 43). Kovalevskaya studied privately in the 1870s with Karl Weierstrass, a professor at the University of Berlin, who submitted her research to the University of Göttingen for the Ph.D. (Koblitz 1993: 123).

3. Etzkowitz et al. (1994) have elaborated the notion of "critical mass for women in science." A minority group is said to have achieved "critical mass" in a population when it reaches a size sufficient to "insure its survival from within and effect a transition to an accepted presence"—a proportion of about 15 percent (51).

4. Uta Merzbach (1989) writes that the statement of historical-critical purpose appeared on the masthead of the *Bulletin* until 1931, when it was quietly dropped (642–643, 653–654).

5. Two of the early presidents of the MAA, Florian Cajori and David Eugene Smith, were not trained as research mathematicians at all but made their reputations as scholars, teachers, and popularizers of the history of mathematics (Merzbach 1989). A third, Herbert Ellsworth Slaught (Ph.D., Chicago, 1898), was among the many MAA members who were instrumental in the founding of the National Council of Teachers of Mathematics (NCTM), an organization devoted to precollege mathematics teaching, in 1920 (Bliss 1938; Cairns 1938).

6. The contributions of women are evident on even casual browsing in the early volumes of the *Monthly*. For a discussion of women in the leadership of the MAA, see Green and LaDuke 1989: 386.

7. The Institute was established with a $5 million gift from Louis Bamberger and Caroline Bamberger Fuld, who had sold their New York retail business just before the 1929 stock market crash. For more on the Institute and its founding, see Institute for Advanced Study 1980; Bers 1989; Borel 1989.

8. Short biographies of Mina Rees and Grace Murray Hopper are found in Grinstein and Campbell 1987. For more on Rees, see Albers and Alexanderson 1985: 255–267; Green et al. 1998. On Hopper, see Rossiter 1995: 268–269 and the references there.

9. On the other hand, at universities that failed to secure a V-12 or ASTP contract, enrollment crises threatened the shutdown of academic programs and the dismissal of faculty. In her interview for the present study, Domina Spencer describes just such a crisis at American University in 1943. The only aspect of the V-12 program in which the university was able to participate was a small program for the training of future military chaplains, which had only "a minimal impact on enrollment" (Amy Robertson, American University Library Archives, e-mail communication, December 1998).

10. In her interview for the present study, Barbara Beechler recalls that she served as a teaching assistant in physics while still a sophomore at the University of Iowa in 1946.

11. On the G.I. Bill, women's colleges, and women's enrollments, see Olson 1974 and Rossiter 1995: 31. The G.I. Bill generally hastened coeducation at women's colleges; in a comparatively rare turnabout, however, "Colgate [in New York State] . . . authorized the wives of veterans to attend its formerly all-male classes" (Olson 1974: 35).

12. Despite considerable social pressure to the contrary, women entered the workforce in increasing numbers in the fifties, though not in high-status jobs; see Kaledin 1984: 61–80.

13. In a 1975 survey of 350 women mathematicians, the spouses of nearly half of those who were married were also mathematicians (Luchins and Luchins 1980: 9). In a similar study of mathematicians in Canada, 45.4 percent of women mathematicians were married to mathematicians (Mura 1990: 77).

14. Claudia Henrion (1997) argues persuasively, for example, that there is no clear correlation between youth and mathematical achievement for either men *or* women (110–113).

15. See, for example, interviews with male mathematicians in Albers and Alexanderson 1985 and Albers, Alexanderson, and Reid 1990; the memoirs of Paul Halmos (1985); reminiscences, anecdotes, and informal histories found in Duren, Askey, and Merzbach 1989; personal reminiscences of Richard Courant and Jerzy Neyman (Reid 1976, 1982); and the vivid depiction of mathematical life at Princeton, NYU, and MIT in the 1950s, in the recent biography of John Nash by Sylvia Nasar (1998). The legendary connection between youth and mathematical achievement has been given official sanction through the institution of the Fields Medals, regarded in the mathematical community as "the equivalent of the Nobel Prize for mathematics." Established in 1932 and awarded at four-year intervals, recipients may be no more than forty years of age (Tropp 1976).

Chapter 2: Women Mathematics Ph.D.'s of the 1940s and 1950s

1. Julia Robinson's life and accomplishments have been recounted in several different publications by her sister, Constance Reid (for example, Reid 1996). Both Cathleen Morawetz and Mary Ellen Rudin have been the subject of numerous interviews and profiles in the popular mathematical press (for example, Albers, Alexanderson, and Reid 1990; Henrion 1997; Morrow and Perl 1998).

2. The first woman elected president of the MAA was Dorothy Bernstein (Ph.D., Brown, 1939), who served in 1979–80. The third woman president of the MAA, Deborah Tepper Haimo (Ph.D., Harvard, 1964), was elected to a two-year term beginning in 1993.

3. After Marjorie Lee Browne earned her doctorate in 1950, the next three African American women to earn Ph.D.'s in mathematics did so in 1960, 1961, and 1962. During the years 1965 to 1981, sixteen more African American women earned the degree (Kenschaft 1981). For more on Granville and Browne, see Kenschaft 1981; Grinstein and Campbell 1987; Morrow and Perl 1998.

4. Karen Uhlenbeck, at Chicago for a brief period in the 1980s, was, after Logsdon, the next tenured woman in the department (Henrion 1997).

5. Marie Wurster (Ph.D., Chicago, 1946), who earned bachelor's and master's degrees in mathematics at Bryn Mawr in 1940 and 1943, respectively, reports that Wheeler—herself a Chicago Ph.D.—strongly influenced her decision to go

to Chicago for the doctorate. Moreover, Wurster decided against attending Radcliffe in part because she would be denied a Harvard degree.

6. For more on Neyman and his support of women as colleagues, collaborators, and students, see Reid 1982. Both Neyman and Scott are profiled in Johnson and Kotz 1997. In addition to Julia Robinson, Tarski's female Ph.D. students at Berkeley include Louise Hoy Chin Lim (1948), Wanda Szmielew (1950), and Anne Davis (1953); see Henkin et al. 1974. Esther Seiden, who shifted her mathematical interests from logic to statistics, describes her experiences at Berkeley and her relationship with Tarski, Neyman, Scott, and others in Samuel-Cahn 1992.

7. The terms *Courant Institute* and *Courant's institute* will be used throughout the book to refer to the graduate center in mathematics founded by Courant in the 1930s. The official name of the graduate center has, however, changed several times over the years. For more on the early history of the Courant Institute, see Morawetz 1989; on Courant's life and work, aspects of his personality, and reminiscences of his friends and colleagues, see Reid 1976.

8. Lipman Bers (1914–1993) was the Ph.D. adviser for a total of sixteen women students over the course of his career at Syracuse, NYU, and Columbia. One of his students, Linda Keen (Ph.D., NYU, 1964), has written an appreciative memorial to him (Keen 1994). See also the interview with Bers in Albers, Alexanderson, and Reid 1990. Wilhelm Magnus (1907–1990) supervised the doctoral dissertations of fourteen women at NYU; he is remembered by nine of his sixty-two doctoral students in Birman and Struik 1991.

9. The 1960s and 1970s saw a proliferation in the literature on sex differences in mathematical abilities (for example, Macoby and Jacklin 1974; Fausto-Sterling 1985; Benbow and Stanley 1992). It is interesting to speculate as to whether, under the influence of this literature, women may have been *directed* into fields— discrete mathematics, for example—deemed better suited to the particular talents of their gender. For a provocative discussion of how social perceptions shape women's performance in mathematics, see Eccles and Jacobs 1992.

10. The sociologists Gaye Tuchman and Nina Fortin (1980) have observed similar patterns of gender inclusion and exclusion in the arts and humanities as well:

[W]hen a field or occupation is not socially valued, women and other minorities will populate it heavily. If the field grows in prestige, (white) men may push women (and other minorities) out. Conversely, as a field loses social value . . . men may decamp and leave the field to women. (309)

11. In fact, one of the most blatant cases of discriminatory antinepotism involves Josephine Mitchell (Ph.D., Bryn Mawr, 1942). Regrettably, I was unable to conduct a formal interview with her for this study. As a tenured associate professor at the University of Illinois in the early 1950s, Mitchell married Lowell Schoenfeld (Ph.D., Pennsylvania, 1944), who was at that time an untenured assistant professor in the same department. Citing antinepotism rules, Illinois chose to terminate *her* contract rather than that of her untenured husband (Rossiter 1995: 125).

12. The word *single* here is somewhat misleading, since many of these unmarried women enjoyed long-term partnerships with other women.

13. Included among the sixteen are Marie Wurster, who joined the faculty of Temple University before it had really achieved research university status; Violet Larney, who joined the faculty of the New York State Teachers College at Albany and remained there through its growth and metamorphosis into the State University of New York at Albany; and Anneli Lax, who was tenured at NYU but whose responsibilities were primarily in undergraduate teaching at Washington Square College.

14. Josephine Mitchell (Ph.D., Bryn Mawr, 1942; see note 11) held professorships at *seven* different colleges and universities before she and her husband were offered tenured professorships at Penn State in 1958. She made these frequent moves in an effort to find an institution where her research would be supported and valued (personal communication, September 1994).

15. Julia Robinson provides the most dramatic example of "sudden tenure." She taught in the Berkeley mathematics department—where her husband, Raphael M. Robinson (Ph.D., California/Berkeley, 1935; 1911–1995), had tenure—on an occasional, as-needed basis for nearly thirty years. When, in 1976, she became the first woman mathematician elected to the National Academy of Sciences, she was made a full professor, essentially overnight (Reid 1996).

16. Among the women mathematics Ph.D.'s of the forties and fifties, Katharine Hazard (Ph.D., Chicago, 1940), Jane Cronin Scanlon (Ph.D., Michigan, 1949), Helen Nickerson (Ph.D., Radcliffe, 1949), Joanne Elliott (Ph.D., Cornell, 1950), and Tilla Weinstein (Ph.D., NYU, 1959) have all held tenured faculty positions at Rutgers University.

17. This is a question she raises explicitly in her interview but more guardedly in Larney 1973.

Chapter 3: Family Background and Early Influences

1. Vivian Eberle Spencer (Ph.D., Pennsylvania, 1936; 1907–1980) wrote a dissertation on Chebyshev polynomials under the direction of Shohat.

2. Ruth Ramler Struik (1894–1993) was perhaps the first woman to earn a Ph.D. in mathematics from Charles University in Prague, where she completed a dissertation in geometry under Georg Pick in 1919. Dirk Jan Struik (born 1894), also a geometer (and later an historian of mathematics) earned his Ph.D. at Leiden in 1922. Her parents met at a mathematical conference in Germany in 1921 and were married two years later (Struik 1994; Rowe 1994).

Joan Rosenblatt's parents, Robert Bruce Raup (1888–1976) and Clara Eliot (1896–1976), met when both were graduate students at Columbia University and married in 1924. They earned Ph.D.'s from Columbia—he in psychology, she in economics—in 1926, the same year in which Joan, their first child, was born.

3. While her sister Vivian had no formal schooling until she entered Oberlin, Domina Spencer did attend a Quaker high school in Philadelphia for four years prior to college.

4. Samuel Karlin (Ph.D., Princeton, 1947) has won numerous awards and honors for his work in mathematical analysis, probability and statistics, operations research, and mathematical biology. He was awarded the National Medal of Science in 1989.

Chapter 4: High School and College

1. The mathematician Karl Menger (1994) has described Vienna in the twenties as "an intellectually lively city" where, in addition to a vibrant artistic and cultural life, one could find

[an] unusually large proportion of professional and business people interested in intellectual achievement. Many members of the legal, financial, and business world; publishers and journalists; physicians and engineers took intense interest in the work of scholars of various kinds. They created an intellectual atmosphere which, I have always felt, few cities enjoyed. (9)

2. Karl Menger (Ph.D., Vienna, 1924; 1902–1985) was a member of the celebrated Vienna Circle of mathematicians and philosophers. He took a professorship at the University of Notre Dame in 1937 and moved to the Illinois Institute of Technology in Chicago in 1948, where he remained until his retirement (Menger 1994; Kass 1996).

3. Seven of the women interviewed—Calloway, Schurrer, Lax, Rubin, Gassner, Luchins, and Weinstein—attended New York City public schools for all or part of their precollege education. In contrast to Violet Larney's experiences in Chicago, female students—particularly those with talent or interest in the sciences—seem to have been taken much more seriously in the New York schools.

4. The term *Hutchins College* refers to University of Chicago president Robert Maynard Hutchins, who believed that "the best time in a student's life to acquire a 'general, higher education' is during the four years between ages sixteen and twenty" (Ward 1992: 82).

5. Born in the Netherlands, Arnold Dresden (Ph.D., Chicago, 1909; 1882–1954) came to the United States in 1903 and taught at Swarthmore from 1927 until his death. A naturalized citizen of the United States, Dresden was a strong supporter of émigré mathematicians during the thirties and forties (Dresden 1942) and a good friend to women in mathematics throughout his career.

6. Larew and Wiggin did their doctoral work in calculus of variations with Gilbert Ames Bliss. When Anderson went to Chicago for the Ph.D., she did her graduate work in the same field with Magnus Hestenes, a student of Bliss. I am grateful to Jeanne LaDuke and Judy Green for information about the pre-1940s Ph.D.'s mentioned in this section.

7. Carlson was the first woman to earn a Ph.D. in mathematics at the University of Minnesota. She worked under the direction of Dunham Jackson, who was later Martin's adviser.

8. Born and educated in Ireland, the mathematician John Lighton Synge (1897–1995) was a prolific researcher in geometry and mathematical physics and held a succession of prestigious university positions in the United States, Canada, and Ireland (Florides 1996). Cecilia Krieger was the first woman to earn a Ph.D. in mathematics at a Canadian university (Anand and Anand 1990).

9. Margaret Young Woodbridge (Ph.D., NYU, 1946) was a doctoral student of Courant; she had previously earned a J.D. from NYU in 1931. Borofsky, Prenowitz, Richardson, and Boyer all earned Ph.D.'s at Columbia in the 1930s and joined the staff of the newly established Brooklyn College soon afterward. Singer earned his Ph.D. at Princeton in 1931.

10. Hubert Evelyn Bray (Ph.D., Rice, 1918) was a Rice Ph.D. and headed the department there for many years. Szolem Mandelbrojt (1899–1983), born into a Jewish family in Poland, moved to France as a child. He earned a Sc.D. in mathematics at the University of Paris in 1923 and was among the founding members of the French mathematics collective, Bourbaki. Mandelbrojt had a lengthy association with Rice and was a full-time member of the department there during World War II. His nephew, Benoit Mandelbrot, is also a mathematician who has gained fame for his work with fractals (Dresden 1942: 428; Corry 1996: 295–304).

11. For more on William LeRoy Hart (Ph.D., Chicago, 1916; 1892–1984), see Price 1986. On Dunham Jackson (Ph.D., Göttingen, 1911; 1888–1946), see Hart 1948. Warren Simms Loud (Ph.D., MIT, 1947; born 1921) earned all of his university degrees at MIT and spent his entire career on the faculty at Minnesota. Neal McCoy's support of women in mathematics is due, in no small part, to the fact that his older sister is also a professional mathematician. Both Neal (born 1905) and Dorothy McCoy (born 1903) earned Ph.D.'s in mathematics under E.W. Chittenden at the University of Iowa in 1929 (see note 14).

12. There have been numerous studies of the life and work of R.L. Moore (Ph.D., Chicago, 1905; 1882–1974). See, in particular, Traylor, Bane, and Jones 1972; Wilder 1976.

13. Students who were not particularly confident to begin with normally did *not* do well with Moore. Although Moore was a teacher whose style suited *her* temperament perfectly, Rudin is quick to add:

I wouldn't for anything have let my children go to school with Moore! That is, I think that he was destructive to anyone who didn't fit exactly into his pattern; he did not succeed in giving the people that worked with him an education. It's a mistake to go to school under those circumstances in general. (Rudin: 22)

As a rule, Moore's students (such as Wilder at Michigan and Rudin herself) were far more compassionate practitioners of the Moore method than Moore himself.

14. E.W. Chittenden (Ph.D., Chicago, 1912; 1886–1977) did research on the boundary between analysis and general topology. In 1918 he joined the faculty at Iowa, where he remained for thirty-six years. Vital and vigorous well into old

age, he worked in a federal laboratory for nine years after his retirement from Iowa and died at the age of ninety-one (Aull 1981).

Gustav Bergmann (Ph.D., Vienna, 1928; 1906–1987), like Karl Menger, had a doctorate in mathematics from Vienna and was a member of the Vienna Circle (see note 2). In 1935 he earned a law degree and practiced law until 1938. In 1939 he left Austria and joined the faculty in philosophy at the University of Iowa, where he remained until his death (Dresden 1942: 423; Brown, Collinson, and Wilkinson 1996: 65).

15. Before earning his Ph.D. in mathematics, Jean van Heijenoort (Ph.D., NYU, 1949; 1912–1986) was active in radical politics and served for a time as Leon Trotsky's bodyguard (Feferman 1993).

The William Lowell Putnam Mathematics Competition, endowed by a wealthy and well-connected Harvard family, is a competitive exam administered annually (with a brief hiatus during World War II) since 1938 by the MAA. The contest is open to individuals and to three-member teams from undergraduate colleges and universities in Canada and the United States. The problems on the examination are fantastically difficult. Monetary prizes are awarded to the top-ranking individuals and teams, and many of the top finishers also win graduate fellowships to Harvard (Birkhoff 1965; Bush 1965; Nasar 1998: 43–44).

16. Every year the *American Mathematical Monthly* reports the results of the Putnam exam. According to the *Monthly,* NYU did not field a Putnam team in 1954 or 1956, but the 1955 Putnam team—consisting of Donald Fredkin, Charles Kahane, and Tilla Weinstein (then Tilla Klotz)—placed among the top eight teams and received honorable mention. No one on the NYU team placed among the top fifteen individuals, so it seems likely that Weinstein and her teammates all performed with roughly equal distinction.

17. For more on Lipman Bers, see chapter 2, especially note 8.

18. The Works Progress Administration (WPA) was a federal jobs program begun in 1935 as part of the New Deal. During its eight years of operation, the WPA created millions of new jobs in construction, education, and the arts (Badger 1989). Max Wertheimer (1880–1943) was born in Czechoslovakia, educated there and in Germany, and came to the New School in New York in 1933. He is best known as the founder of Gestalt psychology but is highly regarded in philosophical circles. His interest in science and mathematics formed the basis of his long association with Albert Einstein (Brown, Collinson, and Wilkinson 1996: 832–833; Kohler 1944).

Chapter 5: Graduate School and the Pursuit of the Ph.D.

1. For a discussion of comparative marriage rates during the Depression, World War II, and the early postwar years, see Hartmann 1982: 164–165; for a thoroughgoing analysis of the interplay of marriage, work, and family in the 1930s, see Scharf 1980.

2. Beginning in the spring of 1944 and continuing into the early postwar years, "graduate work in mathematics was carried on only by women students, men classified as IV-F, and men in war industry who carried on graduate work in the evening" (Kline 1946: 123).

3. Oxtoby and Maharam Stone subsequently became very good friends and colleagues; Oxtoby (1989) wrote a tribute to Maharam Stone and her husband on the occasion of their retirement from the University of Rochester.

4. Katharine Hazard died in the early 1990s, just before the interviewing phase of this project was begun. She had bachelor's and master's degrees from Purdue University, where her father, C.T. Hazard, was on the mathematics faculty (but did not have a doctorate). I was unable to schedule a formal interview with Janet McDonald, although we spoke informally by telephone in December 1994, and she supplied me with written information about her mathematical life and work. She earned a master's degree from Tulane University in 1929 and came to Chicago after many years of small college teaching and administration in her home state of Mississippi. I also spoke informally by telephone with Florence Jacobson during the spring of 1995, not long before her death. A native of Chicago, she began graduate study at the University immediately on the completion of her bachelor's degree there. While she never completed her Ph.D., she went on to a distinguished teaching career at Albertus Magnus College in New Haven, Connecticut.

5. All four of the Chicago advisers had themselves earned Ph.D.'s at Chicago: Graves (Ph.D., 1924; 1896–1973?) and Hestenes (Ph.D., 1932; 1906–1991) were students of G.A. Bliss; Lane (Ph.D., 1918; 1886–1969) was a student of C.J. Wilczynski; Albert (Ph.D., 1928; 1905–1972) wrote a dissertation with L.E. Dickson. See MacLane 1989.

When Florence Dorfman was already fairly far along in her dissertation research, she met her future husband, Nathan Jacobson, when he came to a conference at Chicago. They married in 1942, and she moved with him to the University of North Carolina without completing the Ph.D. Some years later, she did publish the main results of her thesis in a paper written jointly with her husband (Jacobson and Jacobson 1949).

6. Anderson was the only Southern woman among the doctoral students at Chicago in the early forties who did *not* work with Lane.

7. Lane's brand of geometry was regarded with disrespect, perhaps even with derision, by some of his contemporaries. "Lane and his students carried on the study of projective differential geometry using rather crude analytical methods," recalls W.L. Duren (Ph.D., Chicago, 1929). "We who were not Lane's students tended to look on it with disdain." At the same time, Duren adds, "Lane was honest about the shortcomings of the methods, though he did not know how to overcome them" (Duren and Huston 1989: 179).

8. For an especially vivid description of teatime at Princeton in the early 1950s, see Nasar 1998: 63–65. Saunders MacLane lists teatime as among the indispensable ingredients of a "great department of mathematics" (MacLane 1989: 149).

9. Reinhold Baer (Ph.D., Göttingen, 1925; 1902–1979) emigrated from Germany in 1933 and obtained a permanent position at Illinois in 1938. He returned to Germany in 1956 and in the mid-1960s moved to Zurich, where he remained until his death (Gruenberg 1981). Christine Williams Ayoub (Ph.D., Yale, 1947) and Aileen Hostinsky (Ph.D., Illinois, 1949) were among his other female doctoral students.

10. For more about Warren Ambrose (1914–1995), see Halmos 1985; Singer and Wu 1996.

11. Hans Rademacher (Ph.D., Göttingen, 1916; 1892–1969) emigrated to the United States from Germany in 1934, joining the faculty of the University of Pennsylvania the following year. His twenty-one doctoral students include four women, one in Germany and the other three—Ruth Goodman (Ph.D., 1944), Walton, and Leila Dragonette Bram (Ph.D., 1951)—at Penn. Walton experienced Rademacher as imperious and demanding, but reminiscences of Rademacher by his male colleagues (for example, Berndt 1994) praise his talents as a mathematician, teacher, and humanitarian. One such reminiscence, however, provides an illuminating insight: Rademacher insisted that his wife learn to drive so that he would not have to be bothered with doing so himself (Niven 1989: 217).

12. In fact, Wisconsin awarded no Ph.D.'s to women in mathematics during the 1940s. Caroline Lester earned her Ph.D. at Wisconsin under MacDuffee in 1937; it would be thirteen years before Violet Larney completed her Ph.D. with the same adviser in 1950. In addition to Larney and Schurrer, only one other woman, Anna Chandapillai (1954), would earn a Wisconsin mathematics Ph.D. in the 1950s.

13. Jeanne LaDuke and Judy Green have verified that Dorothy McCoy was the only woman to earn a Ph.D. in mathematics at Iowa prior to 1940. The Ph.D. students of Chittenden are listed in Aull 1981.

14. Rebekka Struik's father, Dirk Struik, also fell afoul of the authorities during the McCarthy era (Schrecker 1986; Rowe 1994).

15. Hans Freistadt had political problems of his own. At the University of North Carolina in the 1940s, he did graduate work in physics with fellowship support from the Atomic Energy Commission. At the same time, he was active in the Communist Party. When his political activities came to light, he lost the fellowship in a politically charged, high-profile case (Schrecker 1986: 289–290).

16. In fact, Barrett was filling the vacancy created by the departure of Josephine Mitchell, who had accepted a job at what is now Oklahoma State University. See chapter 2, notes 11 and 14.

17. A total of fifty students completed Ph.D.'s with R.L. Moore from 1916 through 1969; six of them were women. The first of these, Anna Mullikin, received her Ph.D. from Pennsylvania in 1922, although she moved with him to Texas in 1919 and actually completed her studies there (Traylor, Bane, and Jones 1972: 89–90, 196). The other five, all of whom earned Ph.D.'s at Texas, are Harlan Miller (1941), Mary Ellen Rudin (1949), Mary-Elizabeth Hamstrom (1952), Blanche Baker (1965), and Nell Stevenson (1969).

18. J.R. Kline (Ph.D., Pennsylvania, 1916; 1891–1955) was R.L. Moore's first doctoral student. The transfer of Lida Barrett from Moore to Kline was just one example of an established procedure between Moore and his earliest Ph.D. students:

There [developed], across the years, an association between Moore at Texas, Kline at Pennsylvania, Whyburn at Virginia, as well as others which would allow students to move from one place to another, either during their graduate study or following the formal completion of it. . . . Moore or one of his early doctoral students would describe to another the student's capability by naming a certain difficult theorem he had proved. (Traylor, Bane, and Jones 1972: 103)

19. The McCarran-Walter Act (the Immigration and Nationality Act of 1952) was passed in June 1952 over President Truman's veto. It was one of several pieces of legislation from the late 1940s and early 1950s that imposed severe immigration restrictions in an effort to contain the Communist threat from abroad (Loescher and Scanlan 1986).

20. Courant's failure to give proper credit to coauthors—whether male or female—is legendary. The dispute between Courant and Herbert Robbins over the authorship of *What Is Mathematics?* has been debated for years in the mathematical press (see Reid 1976).

Chapter 6: Interweaving a Career and a Life

1. On research opportunities for women in mathematics in the twenties and thirties, see Green and LaDuke 1989: 386–387. Women were regularly on the roster of visiting members in mathematics at the Institute from its inception (Institute for Advanced Study 1980). Margaret Rossiter (1982: 22–24) argues that the women's colleges were themselves complicit in bringing women's research careers in science to an early end.

2. Indeed, with the exception of Marie Wurster, *all* of the women who earned Ph.D.'s in mathematics at Chicago in the 1940s landed their first academic jobs through the old-boy/old-girl network.

3. Dorothy Manning came from an unusual mathematical family with strong ties to Stanford University. Her father, W.A. Manning (Ph.D., Stanford, 1904), earned the first Ph.D. in mathematics ever awarded at Stanford and remained on the faculty there for several decades (Royden 1989: 241–242). Indeed, W.A. Manning may well have been Dorothy Manning's adviser. Her younger sister, Rhoda Manning Wood (Ph.D., Stanford, 1941) completed her Ph.D. at roughly the same time as Clement and thus would have been a natural candidate to succeed her sister at Wells, but she went instead to a position at Oregon State College in the fall of 1941.

Dorothy Manning had an NRC postdoc at Chicago in 1937–38, and spent 1938–39 at the Institute in Princeton before joining the faculty at Wells, but held no further professional positions after her marriage in 1941. She did, however, coauthor two papers with her husband in the early 1960s. Malcolm Smiley

(1912–1982) was Bob Blair's dissertation adviser at Iowa (chapter 5). Smiley's personal experience—having a mathematician wife who gave up her professional career when she married—no doubt affected his interactions with Blair and Barbara Beechler at Iowa.

As in previous chapters, I am indebted to Jeanne LaDuke and Judy Green for detailed information regarding the pre–1940 Ph.D.'s.

4. The mathematics faculty at Vassar normally consisted of at least one faculty member in analysis, one in geometry, and one in algebra. When Asprey joined the Vassar mathematics department in 1945, the other tenure-line faculty—all of them Chicago Ph.D.'s—were Abba Newton (Ph.D., 1933) and Janet McDonald (Ph.D., 1943), both students of E.P. Lane, in geometry; Frances Baker (Ph.D., 1934), a student of Dickson, in algebra; and Wells (Ph.D., 1915), a student of E.H. Moore, in general analysis. Although Asprey was the lone non-Chicago Ph.D., she did have a Chicago connection: her adviser, E.W. Chittenden, had earned his Ph.D. at Chicago under E.H. Moore just three years before Wells (Aull 1981).

5. At the University of Miami, the postwar enrollment surge was particularly dramatic. In the fall of 1945, enrollment at Miami stood at two thousand students. In the fall of 1947, Mary Dean Clement was among *seventy-three* new University of Miami faculty hired to teach a student population of just over eight thousand (Tebeau 1976).

6. The Bateman Manuscript Project, funded by the Office of Naval Research, was a compilation of integral transforms and special functions that built on the work of Harry Bateman (Ph.D., Johns Hopkins, 1913; 1882–1946), a British-born mathematician who spent much of his career at Caltech (Askey 1989). The five volumes associated to the project were published during the years 1953 to 1955. Arthur Erdelyi was the project's director; Wilhelm Magnus, Fritz Oberhettinger, and Francesco Tricomi served as research associates; David Bertin, W.B. Fulks, A.R. Harvey, D.L. Thomsen, Jr., Maria A. Weber, and E.L. Whitney were listed at the front of each volume as research assistants.

7. Paul Franklin Douglass (1904–1988) was a man of diverse accomplishments. As a young man he worked for a time as a newspaper reporter, earned a Ph.D. in political science from the University of Cincinnati (in 1931), worked as an attorney, and won election to the Vermont state legislature while serving as a Methodist pastor. His first real academic appointment was as president of American University, a position he held from 1941 to 1951 (*Who Was Who in America*, volume 9, *s.v.* "Douglass, Paul"). For more on the wartime enrollment crisis at American University, see chapter 1, note 9.

8. Leonard Carmichael, president of Tufts College from 1939 to 1952, sought to transform Tufts into a research institution (Freeland 1992: 179–185). As Spencer relates it, Carmichael wanted to replace the physics department chair with someone more sympathetic to his goals. Spencer found herself on the wrong side of Carmichael when she sided with the chair, who had hired her. Apparently Carmichael believed that having too many women on the faculty in a given department would be "detrimental to its reputation" (Rossiter 1995: 36).

9. Harshbarger 1976 and Eisenhart 1989 provide an overview of the statistical offerings at American colleges and universities before the end of World War II. Because of the interdisciplinary nature of applied statistics, it is not uncommon even today—when most major universities have degree-granting departments of statistics—for statistical research and graduate training to be dispersed among many different departments, such as biostatistics, public health, medicine, or agriculture.

10. Churchill Eisenhart (Ph.D., University College London, 1937; 1913–1994), a Ph.D. student of Jerzy Neyman, worked at NBS from 1947 until 1983, where he served as founding director of the Statistical Engineering Laboratory. His father, Luther Pfahler Eisenhart (Ph.D., Johns Hopkins, 1900; 1876–1965) was also a mathematician—a differential geometer with a secondary interest in statistics (Eisenhart 1989; Cameron and Rosenblatt 1995; Stigler 1995).

11. Lee Lorch (Ph.D., Cincinnati, 1941; born 1915) chaired the mathematics department at Fisk from 1950 to 1955 and was regarded warmly by both faculty and students there (Mayes 1976). During the McCarthy era, Lorch lost a succession of academic positions because of his civil rights activities (Schrecker 1986).

12. For more on the civil rights movement during the Roosevelt administration and increased employment opportunities for blacks in World War II and after, see Perrett 1973, especially chapters 11 and 27.

13. From 1949 to 1950, Cronin Scanlon was briefly married to her dissertation director, Erich Röthe. She states simply, "The marriage did not work at all" (Cronin Scanlon: 18).

14. Paul Erdös (1913–1996), perhaps the most prolific mathematician of the twentieth century, earned a Ph.D. from Cardinal Pazmany University in Budapest, Hungary, in 1934. He was especially well known for posing interesting, difficult, and challenging problems to other mathematicians (Hoffman 1998).

15. For more on the events leading to the passage of Title IX, see Rossi and Calderwood 1973; Sandler 1973; Rossiter 1995. Rossiter writes that "despite the antinepotism rules that had forced some off the faculty in the late 1940s," the University of Wisconsin Medical School employed a husband-and-wife team of cancer researchers, Elizabeth Miller and James Miller, in tenured positions as early as 1959 (1995: 140). In her interview, Mary Ellen Rudin refers to antinepotism at Wisconsin as a generally accepted "practice" rather than a "rule" and says that her promotion came about when the University realized that these practices were being applied inconsistently across departments (Rudin: 18–19).

16. Indeed, during the years 1950 to 1956, Cronin Scanlon published a total of eight papers.

17. Brooklyn Polytechnic has been mentioned several times in this chapter. It was the school that interviewed but did not hire Evelyn Granville after her NYU postdoc and that hired Herbert Morawetz as a professor of chemistry.

18. Kenneth Wolfson (Ph.D., Illinois, 1952; born 1924) spent his entire mathematical career on the faculty of Rutgers University, chairing the mathematics de-

partment there from 1961 to 1975 and serving as dean of the graduate school from 1975 to 1985. In his history of the Rutgers mathematics department, Charles Weibel writes that by 1965, "Rutgers had practically cornered the market on women mathematicians. . . . Seven of the thirty-two senior faculty (22%) were women. The national average was under 1% women then, and is still only 8% today" (Weibel 1995: 12).

19. Indeed, Walter Rudin and Mary Ellen Rudin have written just one joint paper (Rudin and Rudin 1995).

20. Among the mathematicians hired during Tilla Weinstein's tenure as chair at Douglass were Amy Cohen (Ph.D., California/Berkeley, 1970), Jean Taylor (Ph.D., Princeton, 1973), and James Lepowsky (Ph.D., MIT, 1971).

21. Weinstein served a second term as mathematics chair at Douglass from 1978 until the merger of the Douglass and Rutgers departments in 1981.

22. The Mansfield Amendment was enacted in 1969 and remained in effect for just one year. Despite its limited scope, however, the law has been broadly interpreted as sharply curtailing the authority of the Defense Department, and many other federal agencies as well, "to fund basic scientific work that cannot be clearly related to their missions" (David 1985: 45).

23. Eugene Guth (Ph.D., Vienna, 1928; 1905–1990) was a physicist on the faculty of the University of Vienna until 1937, when he became the first Research Professor of Physics at Notre Dame University (Schweinler et al. 1991). Arnold Ross (Ph.D., Chicago, 1931; born 1906), a doctoral student of L.E. Dickson, devoted the early years of his career to research in number theory. It is perhaps ironic, in light of Willerding's observations, that Ross is best known not for his research but for his work with high school teachers and mathematically talented high school students (Lax and Woods 1986; Shapiro 1996).

24. For more on the attitude toward women at Tufts, see note 8.

Chapter 7: Research, Teaching, and the Question of Identity

1. Granville retired from Cal State L.A. in 1984, Texas College in 1987, and the University of Texas at Tyler in 1997.

2. Larney hints at these views in a provocative *Monthly* article (1973), which appeared just as the women's movement was beginning to have an impact on the mathematical community. For a careful analysis of gender differences in research productivity that substantially refutes Larney's claims, see Valian 1998 (especially chapter 12 and the references given there).

3. Harriet Lord, personal communication, October 1996.

4. Wurster asserts repeatedly during her interview that it was her failure to do research, rather than gender discrimination, that caused the difference in pay. But a reward system that values research more highly than textbook publishing and externally funded teaching and service activities is likely to be inherently biased against the sort of activities at which many women excel.

5. Bates's work with Neyman is briefly described in Reid 1982: 227.

6. Edward Kasner (Ph.D., Columbia, 1899; 1878–1955), who spent his professional career at Columbia, was a prolific researcher in differential geometry and mathematical physics. A man of diverse intellectual interests, he gave popular lectures on mathematics in a wide variety of venues, ranging from elementary school classrooms to the New School for Social Research (Douglas 1958).

7. Vera Kistiakowsky (Ph.D., California, 1952: chemistry) and Elizabeth Urey Baranger (Ph.D., Cornell, 1954: physics)—each a daughter of a prominent nuclear scientist—were research associates in physics at MIT in the late 1960s; Vera Pless had become acquainted with them through her husband. In 1972, Kistiakowsky received a sudden promotion to a tenured professorship in physics at MIT, where she remains to this day. Baranger, a Swarthmore College classmate of Rebekka Struik, is currently professor of physics and associate provost at the University of Pittsburgh.

8. Blum 1991 gives a brief history of the first twenty years of AWM.

9. The Committee's first report to the AMS Council is Morawetz 1973.

10. In her interview, Barrett reveals that Anna Pell Wheeler—who remained active in the Philadelphia-area mathematics community on her retirement from Bryn Mawr—played a key role in bringing Barrett to the attention of Wellesley and Bryn Mawr Colleges. See also Grinstein and Campbell 1982.

11. Violet Larney, of course, had held the position of acting chair at SUNY Albany during the academic year 1966–67.

12. See chapter 2, note 2.

Chapter 8: Dimensions of Personal and Professional Success

1. The quote is from Stewart and Vandewater 1993: 253. In the psychologist Erik Erikson's model of adult development, generativity is one of the key tasks at midlife, once concerns with identity and intimacy have been worked out. For a discussion of generativity in older adulthood, see Erikson, Erikson, and Kivnick 1986.

References

Ahern, Nancy C., and Elizabeth L. Scott. 1981. *Career outcomes in a matched sample of men and women Ph.D.'s: An analytical report.* Washington, D.C.: National Academy Press.

Aisenberg, Nadya, and Mona Harrington. 1988. *Women of academe: Outsiders in the sacred grove.* Amherst: University of Massachusetts Press.

Albers, Donald J. 1992. Once upon a time: Anneli Lax and the New Mathematical Library. *MAA Focus* (June): 30–32.

Albers, Donald J., and Gerald L. Alexanderson, editors. 1985. *Mathematical people.* Boston: Birkhauser.

Albers, Donald J., Gerald L. Alexanderson, and Constance Reid, editors. 1990. *More mathematical people: Contemporary conversations.* San Diego: Academic Press.

Anand, Kailash K., and Anita K. Anand. 1990. Cypra Cecilia Krieger and the human side of mathematics. In *Despite the odds: Essays on Canadian women and science,* edited by Marianne Gosztonyi Ainley. Montreal: Vehicule Press.

Askey, Richard. 1989. Handbooks of special functions. In *A century of mathematics in America,* Part 3, edited by Peter Duren, Richard A. Askey, and Uta C. Merzbach. Providence: American Mathematical Society.

Aull, C.E. 1981. E.W. Chittenden and the early history of general topology. *Topology and its applications* 12: 115–125.

Badger, Anthony J. 1989. *The New Deal: The Depression years, 1933–1940.* New York: Farrar, Straus, and Giroux.

Bateson, Mary Catherine. 1990. *Composing a life.* New York: Atlantic Monthly Press, 1989. Reprint, New York: Plume.

Benbow, Camilla Persson, and Julian C. Stanley. 1992. Sex differences in mathematical ability: Fact or artifact? In *The psychology of gender,* Volume 2, edited by Carol Nagy Jacklin. New York: New York University Press. First published in *Science* 210 (1980): 1262–1264.

Berndt, Bruce C. 1994. Hans Rademacher 1892–1969. In *The Rademacher legacy to mathematics,* edited by George E. Andrews, David M. Bressoud, and L. Alayne Parson. Providence: American Mathematical Society.

Bers, Lipman. 1989. The migration of European mathematicians to America. In *A century of mathematics in America,* Part 1, edited by Peter Duren, Richard A. Askey, and Uta C. Merzbach. Providence: American Mathematical Society.

Birkhoff, Garrett. 1965. The William Lowell Putnam Mathematical Competition: Early history. *American Mathematical Monthly* 72: 469–473.

Birman, Joan, and Ruth Rebekka Struik, compilers. 1991. In memoriam: A tribute to Wilhelm Magnus. *Newsletter of the Association for Women in Mathematics* 21 (July/August): 9–14.

Bliss, Gilbert Ames. 1938. Herbert Ellsworth Slaught: Teacher and friend. *American Mathematical Monthly* 45: 5–10.

Blum, Lenore. 1991. A brief history of the Association for Women in Mathematics: The presidents' perspectives. *Notices of the American Mathematical Society* 38: 738–754.

Borel, Armand. 1989. The School of Mathematics at the Institute for Advanced Study. In *A century of mathematics in America,* Part 3, edited by Peter Duren, Richard A. Askey, and Uta C. Merzbach. Providence: American Mathematical Society.

Borland, Katherine. 1991. "That's not what I said": Interpretive conflict in oral narrative research. In *Women's words: The feminist practice of oral history,* edited by Sherna Berner Gluck and Daphne Patai. New York: Routledge.

Boyer, Ernest L. 1990. *Scholarship reconsidered: Priorities of the professoriate.* Princeton: Carnegie Foundation for the Advancement of Teaching.

Brink, Raymond W. 1944. College mathematics during reconstruction. *American Mathematical Monthly* 51: 61–74.

Brown, Dorothy M. 1987. *Setting a course: American women in the 1920s.* Boston: Twayne.

Brown, Ezra A., translator. 1990. *Regiomontanus: His life and work,* by Ernst Zinner. Amsterdam: North Holland.

Brown, Stuart, Diane Collinson, and Robert Wilkinson, editors. 1996. *Biographical dictionary of twentieth-century philosophers.* London: Routledge.

Bush, L.E. 1965. The William Lowell Putnam Competition: Later history and summary of results. *American Mathematical Monthly* 72: 474–483.

Cairns, W.D. 1938. Herbert Ellsworth Slaught: Editor and organizer. *American Mathematical Monthly* 45: 1–4.

Cairns, W.D., Arnold Dresden, and J.R. Kline. 1943. The problem of securing teachers of collegiate mathematics for wartime needs. *Bulletin of the American Mathematical Society* 49: 175–177.

Cameron, Joseph M., and Joan R. Rosenblatt. 1995. Churchill Eisenhart, 1913–1994. *American Statistician* 49: 243–244.

Clowse, Barbara Barksdale. 1981. *Brainpower for the Cold War: The Sputnik crisis and the National Defense Education Act of 1958.* Westport, Conn.: Greenwood Press.

Corry, Leo. 1996. *Modern algebra and the rise of mathematical structures.* Basel: Birkhauser Verlag.

David, Edward E., Jr. 1985. The federal support of mathematics. *Scientific American* 232 (May): 45–51.

Deakin, Michael A.B. 1994. Hypatia and her mathematics. *American Mathematical Monthly* 101: 234–243.

Dolan, Eleanor F., and Margaret P. Davis. 1960. Antinepotism rules in American colleges and universities: Their effect on the faculty employment of women. *The Educational Record* 41: 285–295.

Douglas, Jesse. 1958. Edward Kasner, April 2, 1878–January 7, 1955. In *Biographical memoirs of the National Academy of Sciences,* Volume 31. New York: Columbia University Press.

Dresden, Arnold. 1942. The migration of mathematicians. *American Mathematical Monthly* 49: 415–429.

Duncan, Erika. 1989. On preserving the union of numbers and words: The story of an experiment. In *Writing to learn mathematics and science,* edited by Paul Connolly and Theresa Vilardi. New York: Teachers College Press.

Duren, Peter, Richard A. Askey, and Uta C. Merzbach, editors. 1989. *A century of mathematics in America.* Parts 1–3. Providence: American Mathematical Society.

Duren, William L., Jr., with Antoinette Killen Huston. 1989. Graduate student at Chicago in the twenties. In *A century of mathematics in America,* Part 2, edited by Peter Duren, Richard A. Askey, and Uta C. Merzbach. Providence: American Mathematical Society. First published in *American Mathematical Monthly* 83 (1976): 243–247.

Dzielska, Maria. 1995. *Hypatia of Alexandria.* Translated by F. Lyra. Cambridge: Harvard University Press.

Eccles, Jacquelynne S., and Janis E. Jacobs. 1992. Social forces shape math attitudes and performance. In *The psychology of gender,* Volume 2, edited by Carol Nagy Jacklin. New York: New York University Press. First published in *Signs: Journal of Women in Culture and Society* 11 (1986): 367–380.

Eisenhart, Churchill. 1989. S.S. Wilks' Princeton appointment, and statistics at Princeton before Wilks. In *A century of mathematics in America,* Part 3, edited by Peter Duren, Richard A. Askey, and Uta C. Merzbach. Providence: American Mathematical Society.

Erikson, Erik H., Joan M. Erikson, and Helen Q. Kivnick. 1986. *Vital involvement in old age.* New York: Norton.

Etzkowitz, Henry, Carol Kemelgor, Michael Neuschatz, Brian Uzzi, and Joseph Alonzo. 1994. The paradox of critical mass for women in science. *Science* 266: 51–54.

Ewing, John H., editor. 1994. *A century of mathematics through the eyes of the Monthly.* Washington, D.C.: Mathematical Association of America.

Fausto-Sterling, Anne. 1985. *Myths of gender: Biological theories about men and women.* New York: Basic Books.

Feferman, Anita Burdman. 1993. *Politics, logic, and love: The life of Jean van Heijenoort.* Wellesley, Mass.: Peters.

Fenster, Della Dumbaugh, and Karen Parshall. 1994. Women in the American mathematical research community: 1891–1906. In *The history of modern mathematics.* Volume 3, *Images, ideas, and communities,* edited by E. Knobloch and D. Rowe. San Diego: Academic Press.

Florides, P.S. 1996. Obituary: Professor John Lighton Synge, FRS. *Irish Mathematical Society Bulletin* 37: 3–6.

Fox, Lynn H., Linda Brody, and Dianne Tobin, editors. 1980. *Women and the mathematical mystique.* Baltimore: Johns Hopkins University Press.

Freeland, Richard M. 1992. *Academia's golden age: Universities in Massachusetts, 1945–1970.* Oxford: Oxford University Press.

Freitag, Herta T., and Arthur H. Freitag. 1956. Neopythagorean triangles. *Scripta Mathematica* 22: 122–131.

Friedan, Betty. 1963. *The feminine mystique.* New York: Norton.

Graham, Patricia Albjerg. 1978. Expansion and exclusion: A history of women in American higher education. *Signs: Journal of Women in Culture and Society* 3: 759–773.

Green, Judy, and Jeanne LaDuke. 1987. Women in the American mathematical community: The pre-1940 Ph.D.'s. *Mathematical Intelligencer* 9: 11–23.

Green, Judy, and Jeanne LaDuke. 1989. Women in American mathematics: A century of contributions. In *A century of mathematics in America,* Part 2, edited by Peter Duren, Richard A. Askey, and Uta C. Merzbach. Providence: American Mathematical Society.

Green, Judy, Jeanne LaDuke, Saunders MacLane, and Uta Merzbach. 1998. Mina Spiegel Rees (1902–1997). *Notices of the American Mathematical Society* 45: 866–873.

Grinstein, Louise S., and Paul J. Campbell. 1982. Anna Johnson Pell Wheeler: Her life and work. *Historia Mathematica* 9: 37–53.

Grinstein, Louise S., and Paul J. Campbell, editors. 1987. *Women of mathematics: A biobibliographic sourcebook.* Westport, Conn.: Greenwood Press.

Gruenberg, K.W. 1981. Reinhold Baer. *Bulletin of the London Mathematical Society* 13: 339–361.

Hall, Tord. 1970. *Carl Friedrich Gauss: A biography.* Translated by Albert Froderberg. Cambridge: MIT Press.

Halmos, Paul R. 1985. *I want to be a mathematician.* New York: Springer-Verlag.

Harcleroad, Fred F., and Allan W. Ostar. 1987. *Colleges and universities for change: America's comprehensive public state colleges and universities.* Washington, D.C.: American Association of State Colleges and Universities.

Harmon, Lindsey R., compiler. 1978. *A century of doctorates: Data analyses of growth and change*. Washington, D.C.: National Academy of Sciences/National Research Council.

Harmon, Lindsey R., and Herbert Soldz, compilers. 1963. *Doctorate production in United States universities 1920–1962*. Washington, D.C.: National Academy of Sciences/National Research Council.

Harshbarger, Boyd. 1976. History of the early developments of modern statistics in America (1920–1944). In *On the history of statistics and probability*, edited by D.B. Owen. New York: Marcel Dekker.

Hart, William L. 1948. Dunham Jackson 1888–1946. *Bulletin of the American Mathematical Society* 54: 847–860.

Hartmann, Susan M. 1982. *The home front and beyond: American women in the 1940s*. Boston: Twayne.

Helson, Ravenna. 1967. Sex differences in creative style. *Journal of Personality* 35: 214–233.

Helson, Ravenna. 1968. Effects of sibling characteristics and parental values on creative interest and achievement. *Journal of Personality* 36: 589–607.

Helson, Ravenna. 1971. Women mathematicians and the creative personality. *Journal of Consulting and Clinical Psychology* 36: 210–220.

Henkin, Leon, John Addison, C.C. Chang, William Craig, Dana Scott, and Robert Vaught, editors. 1974. *Proceedings of the Tarski symposium: An international symposium held to honor Alfred Tarski on the occasion of his seventieth birthday*. Proceedings of Symposia in Pure Mathematics, Volume 25. Providence: American Mathematical Society.

Henrion, Claudia A. 1997. *Women in mathematics: The addition of difference*. Bloomington: Indiana University Press.

Herstein, I.N. 1969. On the Ph.D. in mathematics. *American Mathematical Monthly* 76: 818–824.

Hochschild, Arlie Russell. 1975. Behind the clockwork of men's careers. In *Women and the power to change*, edited by Florence Howe. New York: McGraw-Hill.

Hoffman, Paul. 1998. *The man who loved only numbers: The story of Paul Erdös and the search for mathematical truth*. New York: Hyperion.

Horowitz, Helen Lefkowitz. 1987. *Campus life: Undergraduate cultures from the end of the eighteenth century to the present*. Chicago: University of Chicago Press.

Hutchinson, Joan. 1977. "Let me count the ways": Women in combinatorics. *Newsletter of the Association for Women in Mathematics* 7 (January/February): 3–7.

Hutchinson, Joan. 1994. A note on Sister Celine. *Newsletter of the Association for Women in Mathematics* 24 (May/June): 8.

Institute for Advanced Study. 1980. *A community of scholars: The Institute for Advanced Study, faculty and members 1930–1980*. Princeton: Princeton University Press.

Jacklin, Carol Nagy, editor. 1992. *The psychology of gender.* Volumes 1–4. New York: New York University Press.

Jacobson, F.D., and N. Jacobson. 1949. Classification and representation of semi-simple Jordan algebras. *Transactions of the American Mathematical Society* 65: 141–169.

Jewett, Frank B. 1942. The mobilization of science in national defense. *Science* 95: 235–241.

Johnson, Norman L., and Samuel Kotz, editors. 1997. *Leading personalities in the statistical sciences: From the seventeenth century to the present.* New York: Wiley.

Kaledin, Eugenia. 1984. *Mothers and more: American women in the 1950s.* Boston: Twayne.

Kanter, Rosabeth Moss. 1977. *Men and women of the corporation.* New York: Basic Books.

Kaplan, Wilfred. 1989. Mathematics at the University of Michigan. In *A century of mathematics in America,* Part 3, edited by Peter Duren, Richard A. Askey, and Uta C. Merzbach. Providence: American Mathematical Society.

Kass, Seymour. 1996. Karl Menger. *Notices of the American Mathematical Society* 43: 558–561.

Keen, Linda. 1994. Lipman Bers (1914–1993). *Newsletter of the Association for Women in Mathematics* 24 (May/June): 5–7.

Kenschaft, Patricia C. 1981. Black women in mathematics in the United States. *American Mathematical Monthly* 88: 592–604.

King, Amy C. and Rosemary McCroskey. 1976–1977. Women Ph.D.'s in mathematics in USA and Canada: 1886–1973. *Philosophia Mathematica* 13/14: 79–129.

Kleinman, Daniel Lee. 1995. *Politics on the endless frontier: Postwar research policy in the United States.* Durham: Duke University Press.

Kline, J.R. 1946. Rehabilitation of graduate work. *American Mathematical Monthly* 53: 121–131.

Koblitz, Ann Hibner. 1993. *A convergence of lives. Sofia Kovalevskaya: Scientist, writer, revolutionary.* New Brunswick: Rutgers University Press.

Kohler, Wolfgang. 1944. Max Wertheimer 1880–1943. *Psychological Review* 51: 142–146.

Kohlstedt, Sally Gregory. 1978. In from the periphery: American women in science 1830–1880. *Signs: Journal of Women in Culture and Society* 4: 81–96.

Larney, Violet H. 1973. Female mathematicians, where are you? *American Mathematical Monthly* 80: 310–313.

Lax, Anneli. 1989. They think, therefore we are. In *Writing to learn mathematics and science,* edited by Paul Connolly and Theresa Vilardi. New York: Teachers College Press.

Lax, Anneli, and Alan C. Woods. 1986. Award for distinguished service to Professor Arnold Ephraim Ross. *American Mathematical Monthly* 93: 245–246.

Loescher, Gil, and John A. Scanlan. 1986. *Calculated kindness: Refugees and America's half-open door, 1945 to the present.* New York: Free Press.

Luchins, Edith H., and Abraham S. Luchins. 1980. Female mathematicians: A contemporary appraisal. In *Women and the mathematical mystique,* edited by Lynn H. Fox, Linda Brody, and Dianne Tobin. Baltimore: Johns Hopkins University Press.

Maccoby, Eleanor Emmons, and Carol Nagy Jacklin. 1974. *The psychology of sex differences.* Stanford: Stanford University Press.

MacLane, Saunders. 1989. Mathematics at the University of Chicago: A brief history. In *A century of mathematics in America,* Part 2, edited by Peter Duren, Richard A. Askey, and Uta C. Merzbach. Providence: American Mathematical Society.

Mayes, Vivienne. 1976. Lee Lorch at Fisk: A tribute. *American Mathematical Monthly* 83: 708–711.

Menger, Karl. 1994. *Reminiscences of the Vienna Circle and the mathematical colloquium.* Edited by Louise Golland, Brian McGuinness, and Abe Sklar. Dordrecht: Kluwer.

Merzbach, Uta C. 1989. The study of the history of mathematics in America: A centennial sketch. In *A century of mathematics in America,* Part 3, edited by Peter Duren, Richard A. Askey, and Uta C. Merzbach. Providence: American Mathematical Society.

Morawetz, Cathleen. 1973. Women in mathematics. *Notices of the American Mathematical Society* 3: 131–132.

Morawetz, Cathleen. 1989. The Courant Institute of Mathematical Sciences. In *A century of mathematics in America,* Part 2, edited by Peter Duren, Richard A. Askey, and Uta C. Merzbach. Providence: American Mathematical Society.

Morrow, Charlene, and Teri Perl, editors. 1998. *Notable women in mathematics: A biographical dictionary.* Westport, Conn.: Greenwood Press.

Mura, Roberta. 1990. *Profession: Mathématicienne. Étude comparative des professeur-e-s universitaires en sciences mathématiques.* Québec: Groupe de recherche multidisciplinaire féministe, Université Laval.

Nasar, Sylvia. 1998. *A beautiful mind: A biography of John Forbes Nash, Jr.* New York: Simon and Schuster.

National Research Council. 1996. Office of Scientific and Engineering Personnel. Survey of earned doctorates. Washington, D.C.: National Academy of Sciences. Unpublished tables.

Newcomer, Mabel. 1959. *A century of higher education for American women.* New York: Harper.

Newsom, C.V., editor. 1943a. The Army Specialized Training Program. *American Mathematical Monthly* 50: 466–470.

Newsom, C.V., editor. 1943b. The Navy College Training Program. *American Mathematical Monthly* 50: 645–650.

Niven, Ivan. 1989. The threadbare thirties. In *A century of mathematics in America,* Part I, edited by Peter Duren, Richard A. Askey, and Uta C. Merzbach. Providence: American Mathematical Society.

Olson, Keith W. 1974. *The G.I. Bill, the veterans, and the colleges.* Lexington: University Press of Kentucky.

Osen, Lynn. 1974. *Women in mathematics.* Cambridge: MIT Press.

Oxtoby, John C. 1989. Biographical note. In *Measure and measurable dynamics: Proceedings of a conference held in honor of Dorothy Maharam Stone held September 17–19, 1987,* edited by R. Daniel Mauldin, R.M. Shortt, and Cesar E. Silva. Contemporary Mathematics, Volume 94. Providence: American Mathematical Society.

Parshall, Karen V.H., and David E. Rowe. 1994. *The emergence of the American mathematical research community 1876–1900: J.J. Sylvester, Felix Klein, and E.H. Moore.* Providence: American Mathematical Society.

Perrett, Geoffrey. 1973. *Days of sadness, years of triumph: The American people 1939–1945.* New York: Coward, McCann, and Geoghegan.

Power, Edward J. 1972. *Catholic higher education in America.* New York: Appleton-Century-Crofts.

Price, G. Baley. 1943. Adjustments in mathematics to the impact of war. *American Mathematical Monthly* 50: 31–34.

Price, G. Baley. 1986. William LeRoy Hart 1892–1984. *Mathematics Magazine* 59: 232–238.

Rainsford, George N. 1972. *Congress and higher education in the nineteenth century.* Knoxville: University of Tennessee Press.

Rainville, Earl. 1960. *Special functions.* New York: Macmillan.

Rees, Mina. 1989. The mathematical sciences and World War II. In *A century of mathematics in America,* Part 1, edited by Peter Duren, Richard A. Askey, and Uta C. Merzbach. Providence: American Mathematical Society. First published in *American Mathematical Monthly* 87 (1980): 607–621.

Reid, Constance. 1976. *Courant in Göttingen and New York: The story of an improbable mathematician.* New York: Springer-Verlag.

Reid, Constance. 1982. *Neyman—from life.* New York: Springer-Verlag.

Reid, Constance. 1996. *Julia: A life in mathematics.* Washington, D.C.: Mathematical Association of America.

Reingold, Nathan. 1989. Refugee mathematicians in the USA. In *A century of mathematics in America,* Part 1, edited by Peter Duren, Richard A. Askey, and Uta C. Merzbach. Providence: American Mathematical Society. First published in *Annals of Science* 38 (1981): 313–338.

Richardson, R.G.D. 1943. Applied mathematics and the present crisis. *American Mathematical Monthly* 50: 415–423.

Richardson, R.G.D. 1989. The Ph.D. degree and mathematical research. In *A century of mathematics in America*, Part 2, edited by Peter Duren, Richard A. Askey, and Uta C. Merzbach. Providence: American Mathematical Society. First published in *American Mathematical Monthly* 43 (1936): 199–215.

Rossi, Alice S., and Ann Calderwood, editors. 1973. *Academic women on the move*. New York: Russell Sage.

Rossiter, Margaret W. 1978. Sexual segregation in the sciences: Some data and a model. *Signs: Journal of Women in Culture and Society* 4: 146–151.

Rossiter, Margaret W. 1982. *Women scientists in America: Struggles and strategies to 1940*. Baltimore: Johns Hopkins University Press.

Rossiter, Margaret W. 1995. *Women scientists in America: Before affirmative action 1940–72*. Baltimore: Johns Hopkins University Press.

Rowe, David E. 1994. Dirk Jan Struik and his contributions to the history of mathematics. *Historia Mathematica* 21: 245–273.

Royden, Halsey. 1989. A history of mathematics at Stanford. In *A century of mathematics in America*, Part 2, edited by Peter Duren, Richard A. Askey, and Uta C. Merzbach. Providence: American Mathematical Society.

Rudin, Mary Ellen, and Walter Rudin. 1995. Continuous functions that are locally constant on dense sets. *Journal of Functional Analysis* 133: 129–137.

Ruskai, Mary Beth. 1994. Myths about the role of marital status in career advancement. *Newsletter of the Association for Women in Mathematics* 24 (May/June): 9–11.

Samuel-Cahn, Ester. 1992. A conversation with Esther Seiden. *Statistical Science* 7: 339–357.

Sandler, Bernice. 1973. A little help from our government: WEAL and contract compliance. In *Academic women on the move*, edited by Alice S. Rossi and Ann Calderwood. New York: Russell Sage.

Sarton, George. 1936. *The study of the history of mathematics*. Cambridge: Harvard University Press.

Scharf, Lois. 1980. *To work and to wed: Female employment, feminism, and the Great Depression*. Westport, Conn.: Greenwood Press.

Schrecker, Ellen. 1986. *No ivory tower: McCarthyism and the universities*. Oxford: Oxford University Press.

Schweinler, Harold, Burak Erman, James E. Mark, and Alvin Weinberg. 1991. Eugene Guth (obituary). *Physics Today* 44: 133–134.

Shapiro, Daniel B. 1996. A conference honoring Arnold Ross on his ninetieth birthday. *Notices of the American Mathematical Society* 43: 1151–1154.

Simpson, Renate. 1983. *How the Ph.D. came to Britain: A century of struggle for postgraduate education*. Surrey (U.K.): Society for Research into Higher Education.

Singer, I.M., and H. Wu. 1996. A tribute to Warren Ambrose. *Notices of the American Mathematical Society* 43: 425–427.

Solomon, Barbara Miller. 1985. *In the company of educated women.* New Haven: Yale University Press.

Stephan, Paula E., and Mary Mathewes Kassis. 1997. The history of women and couples in academe. In *Academic couples: Problems and promises,* edited by Marianne A. Ferber and Jane W. Loeb. Urbana: University of Illinois Press.

Stewart, Abigail J., and Elizabeth A. Vandewater. 1993. The Radcliffe class of 1964: Career and family social clock projects in a transitional cohort. In *Women's lives through time: Educated women of the twentieth century,* edited by Kathleen Day Hulbert and Diane Tickton Schuster. San Francisco: Jossey-Bass.

Stigler, Stephen M. 1995. Eloge: Churchill Eisenhart, 11 March 1913–25 June 1994. *Isis* 86: 455–456.

Struik, Dirk J. 1989. The MIT department of mathematics during its first seventy-five years: Some recollections. In *A century of mathematics in America,* Part 3, edited by Peter Duren, Richard A. Askey, and Uta C. Merzbach. Providence: American Mathematical Society.

Struik, Dirk J. 1994. Saly Ruth Ramler Struik. Unpublished manuscript.

Sulloway, Frank J. 1996. *Born to rebel.* New York: Pantheon.

Tebeau, Charlton W. 1976. *The University of Miami: A golden anniversary history 1926–1976.* Coral Gables: University of Miami Press.

Tobias, Sheila. 1993. *Overcoming math anxiety.* New York: Norton.

Traylor, D. Reginald, William Bane, and Madeline Jones. 1972. *Creative teaching: Heritage of R.L. Moore.* Houston: University of Houston Press.

Tropp, Henry S. 1976. The origins and history of the Fields medal. *Historia Mathematica* 3: 167–181.

Tuchman, Gaye, and Nina E. Fortin. 1980. Edging women out: Some suggestions about the structure of opportunities and the Victorian novel. *Signs: Journal of Women in Culture and Society* 6: 308–325.

Valian, Virginia. 1998. *Why so slow? The advancement of women.* Cambridge: MIT Press.

Wallenstein, Peter. 1997. *Virginia Tech, land-grant university, 1872–1997: History of a school, a state, a nation.* Blacksburg, Va.: Pocahontas Press.

Ward, F. Champion. 1992. Requiem for the Hutchins College. In *General education in the social sciences: Centennial reflections on the College of the University of Chicago,* edited by John J. MacAloon. Chicago: University of Chicago Press.

Ware, Susan. 1982. *Holding their own: American women in the 1930s.* Boston: Twayne.

Warnick, Mark S. 1994. Five-foot nun a math giant. *Pittsburgh Post-Gazette,* 15 May, C1 and C4.

Weibel, Charles. 1995. A history of mathematics at Rutgers. Unpublished manuscript, available on the Internet at http://www.math.rutgers.edu/~weibel/history.html.

Wilder, Raymond L. 1976. Robert Lee Moore, 1882–1974. *Bulletin of the American Mathematical Society* 82: 417–427.

Young, J.W. 1932. Functions of the Mathematical Association of America. *American Mathematical Monthly* 39: 6–15.

Zachary, G. Pascal. 1997. *Endless frontier: Vannevar Bush, engineer of the American century.* New York: Free Press.

Zeilberger, Doron. 1982. Sister Celine's technique and its generalizations. *Journal of Mathematical Analysis and Its Applications* 85: 114–115.

Photo Credits

Anne Lewis (Anderson) with Magnus Hestenes (left) of UCLA and E.B. Shanks of Vanderbilt University, at the meeting of the Southeastern Section of the MAA in Greensboro, North Carolina, 1962 (Photograph courtesy of Anne Lewis Anderson, reprinted by permission of the University of North Carolina at Greensboro)

Winifred Asprey at the IBM/360, Vassar College, about 1968 (Photograph courtesy of Winifred Asprey)

Lida Barrett, 1950 (Photograph courtesy of Lida and Maidel Barrett)

Grace Bates, 1996 (Photograph taken by the author)

Barbara Beechler, 1996 (Photograph taken by the author)

Jane Cronin Scanlon, 1996 (Photograph taken by the author)

Patricia Eberlein, 1997 (Photograph taken by the author)

Herta Freitag, 1995 (Photograph taken by the author)

Evelyn Granville, 1997 (Photograph taken by the author)

Susan Hahn, 1997 (Photograph taken by the author)

Anneli Lax, 1997 (Photograph taken by the author)

Edith Luchins in the 1990s (Photograph courtesy of Edith Luchins)

Margaret Martin, 1997 (Photograph taken by the author)

Cathleen Morawetz in her office at the Courant Institute, 1983 (Photograph © New York University, reprinted by permission of Cathleen Morawetz and the New York University Photo Bureau)

Vivienne Morley, 1997 (Photograph taken by the author)

Mary Ellen Rudin, about 1990 (Photograph courtesy of Mary Ellen Rudin)

Alice Schafer in the classroom, Wellesley College, 1973 (Photograph taken by Bradford F. Herzog, courtesy of the Wellesley College Office of Public Information)

Augusta Schurrer, 1996 (Photograph taken by the author)

Domina Eberle Spencer and Hypatia in Venice, 1998 (Photograph courtesy of Domina Eberle Spencer)

Maria Steinberg, 1996 (Photograph taken by the author)

Rebekka Struik, 1997 (Photograph taken by the author)

Jean Walton, 1996 (Photograph taken by the author)

Tilla Weinstein (center) with Lipman and Mary Bers, 1991 (Photograph courtesy of Tilla Weinstein)

Index